Concept Book

A Mathematics Reference for Teachers and Students

America's Choice® is a subsidiary of the National Center on Education and the Economy® (NCEE), a Washington, DC-based non-profit organization and a leader in standards-based reform. In the late 1990s, NCEE launched the America's Choice School Design, a comprehensive, standards-based, school-improvement program that serves students through partnerships with states, school districts, and schools nationwide. In addition to the school design, America's Choice provides instructional systems in literacy, mathematics, and school leadership. Consulting services are available to help school leaders build strategies for raising student performance on a large scale.

ISBN 1-932976-44-2

www.americaschoice.org
products@americaschoice.or
800-221-3641

First printing 2005
4 5 6 7 8 9 10 09 08

Table of Contents

Chapter

Table of Contents

Chapter

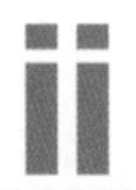

CHAPTER 1

MATHEMATICAL REASONING

Particular Statements and General Statements

Mathematical statements can be true or untrue, and they can be about particular things or about more general classes of things.

Example

The statement $2 + 2 = 2 \bullet 2$ is a true statement about some particular numbers.

The statement $3 + 3 = 3 \bullet 3$ is another statement, but it is untrue.

The statement $x = x + 1$ is a general statement but it is untrue, no matter what value is chosen for x. In words, this untrue statement says, "A number is equal to 1 more than itself." This can never happen.

The statement $a + b = b + a$ is a general statement about adding two numbers a and b. It is a true statement for any numbers a and b. The most important statements in mathematics are of this kind, general and true.

Chapter 1

Always, Sometimes, Never

General statements about numbers in mathematics are either *always true*, *sometimes true*, or *never true.*

Example

The statement $a + b = b + a$ is *always true*. It is the commutative property of addition and holds for all numbers a and b.

The statement $a = a + 1$ is *never true*. It holds for no number a.

The statement $a + a = a \bullet a$ is *sometimes true*. It is true for some values of a, and *never true* for other values.

It is true for 2, since $2 + 2 = 4$, and $2 \bullet 2 = 4$.

It is false for many numbers, such as 3, since $3 + 3 = 6$, but $3 \bullet 3 = 9$.

Examples and Counterexamples

To explain why a general statement is *sometimes true*, you need two things:

An *example* showing a case when it is true, such as $x = 2$ in $x + x = x \bullet x$

A *counterexample* showing a case when it is untrue, such as $x = 3$ in $x + x = x \bullet x$

Example

$x + x = x \bullet x$ is *sometimes true.*

$x = 2$ makes the equation true. This is the example.

$x = 3$ makes the equation false. This is the counterexample.

True and Untrue for Different Types of Numbers

Often the truth of a statement depends on the type of number you are discussing.

Example

When you write about factors and multiples, it is normally assumed that you are writing about whole numbers.

So, even though $\frac{1}{2} \bullet 14 = 7$, you would not say that $\frac{1}{2}$ and 14 are factors of 7, since $\frac{1}{2}$ is not a whole number.

Example

Think about the statement $a \bullet b > a$. If you think about whole numbers and multiplications you might think that this statement is *always true* for all numbers a and b.

But looking at a larger set of numbers, you need to break down the possible values of a and b into all their separate cases. The following table sorts out when the statement $a \bullet b > a$ is true and when it is false.

	$b > 1$	$b = 1$	$b < 1$
$a > 0$	True	False, $a \bullet b = a$	False, $a \bullet b < a$
$a = 0$	False, $a \bullet b = a$	False, $a \bullet b = a$	False, $a \bullet b = a$
$a < 0$	False, $a \bullet b < a$	False, $a \bullet b = a$	True

Using this analysis, the *sometimes true* statement $a \bullet b > a$ can be replaced with these two statements:

$a \bullet b > a$ is true for $a > 0$ and $b > 1$ and, also, for $a < 0$ and $b < 1$.

$a \bullet b \leq a$ is true for $a > 0$ and $b \leq 1$ and, also, for $a < 0$ and $b \geq 1$, and $a = 0$.

Always true statements are more useful to mathematicians than *sometimes true* statements, because *always true* statements help them examine groups of values.

Equations

The truth of an equation is often undecided until the variable is replaced with a chosen value.

Example

$x + x = x \bullet x$

Replacing x with 2 gives $2 + 2 = 2 \bullet 2$, which is a true statement.

Replacing x with 0 gives $0 + 0 = 0 \bullet 0$, which is a true statement.

Replacing x with 3 gives $3 + 3 = 3 \bullet 3$, which is a false statement.

The values of the variable that make the equation a true statement are called the *solutions* to the equation. You can use algebra to show that $x = 0$ and $x = 2$ are the only solutions to the equation $x + x = x \bullet x$.

The equation $x^2 + 2x = x(x + 2)$ follows from the distributive property. It is true for all values of x. An equation true for all values of a variable is called an *identity*.

The equation $x = x + 1$ has no solutions.

Finding Out and Saying Why

In mathematics, you can find interesting number patterns or statements that appear to be true in a large number of cases. Justifying or explaining why these number patterns or statements are true in all cases is also part of mathematics. This mathematical process is called a *justification*.

Example

Finding the Pattern

Look for patterns in the calculations below. Note that many digits are reversed, and that many of the equations involve subtracting or dividing by 9.

$73 - 37 = 36$	$36 \div 9 = 4$
$82 - 28 = 54$	$54 \div 9 = 6$
$75 - 57 = 18$	$18 \div 9 = 2$
$98 - 89 = 9$	$9 \div 9 = 1$

$987 - 789 = 198$	$198 \div 9 = 22$
$786 - 678 = 108$	$108 \div 9 = 12$
$150 - 015 = 135$	$135 \div 9 = 15$
$9{,}753 - 3{,}579 = 6{,}174$	$6{,}174 \div 9 = 686$

Claim: If the difference is calculated between a number and the number with digits reversed, this difference is a number that is divisible by 9.

Saying Why—The Justification

A mathematical justification must go beyond only writing examples. To justify the statement for two-digit numbers, you could write a two-digit number as $10a + b$, where a is the tens digit and b is the ones digit: for example, $73 = 10 \bullet 7 + 3$, where $a = 7$ and $b = 3$. With the digits reversed, the number is $10b + a$: $37 = 10 \bullet 3 + 7$.

Given	$(10a + b)$	the number
	$(10b + a)$	the digits reversed
	$(10a + b) - (10b + a)$	the difference
Then	$= 10a + b - 10b - a$	distributive property
	$= 10a - a - 10b + b$	rearranging terms
	$= 9a - 9b$	collecting like terms
	$= 9(a - b)$	distributive property
	$9(a - b)$	is a multiple of 9

Tools for Mathematical Reasoning

To write a mathematical justification, you can use these tools:

Tools for Justification	
	• Definitions—for example, the definition of an even number • Properties—for example, the distributive property • Previously known results—for example, the sum of angles for a triangle • Given information—a fact included in the problem statement • Diagrams and words—to explain your reasoning • Letters—to represent the variables • Examples and counterexamples—for *sometimes true* statements

A Proof Using Number Properties

An arithmetic sequence is a list of numbers that are an equal distance apart on the number line, such as 92, 96, 100, 104, 108. Each number is 4 more than the previous number.

Example

Claim: The sum of five numbers in an arithmetic sequence is five times the middle number.

Example: For the sequence 92, 96, 100, 104, 108, the sum is $5 \bullet 100 = 500$.

To justify the claim, use letters to represent any set of five numbers and then use the number properties to complete your justification.

Let the first number be a, and let the step size be d. Then the other four numbers are $a + d$, $a + 2d$, $a + 3d$, and $a + 4d$

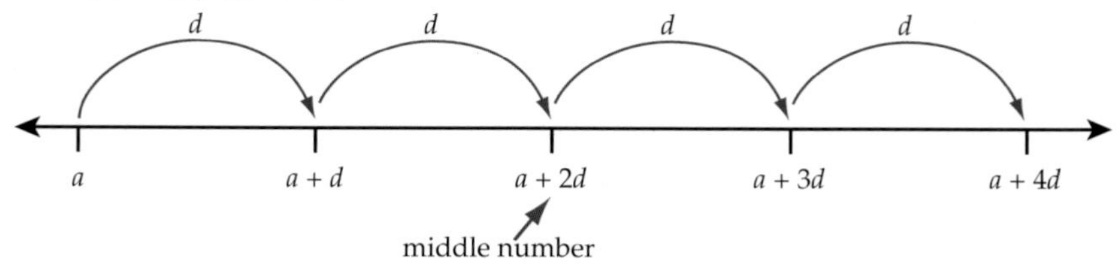

The sum of the five numbers is given by:

$S = a + (a + d) + (a + 2d) + (a + 3d) + (a + 4d)$	
$= a + a + d + a + 2d + a + 3d + a + 4d$	
$= (a + a + a + a + a) + (d + 2d + 3d + 4d)$	regrouping terms
$= 5a + 10d$	
$= 5(a + 2d)$	distributive property
$5(a + 2d)$	which is five times the middle number $a + 2d$

This completes the proof of the claim.

A Proof in Geometry

Consider the statement, "The sum of the measures of the interior angles of a quadrilateral is 360°."

To prove this, you need to know the definition of a quadrilateral:
a quadrilateral is a four-sided polygon.

You also need to know that the sum of the measures of the angles in a triangle is 180°.

You can use a diagram like the one below to explain that one of the diagonals of the quadrilateral will divide the quadrilateral into two triangles. Two of the four angles in the quadrilateral are also divided into smaller angles.

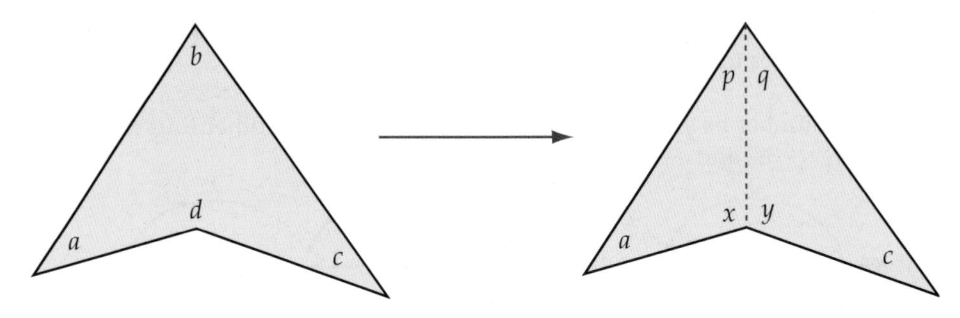

With or without using the letters in the diagram, you can then argue that the sum of the four angles in the quadrilateral is equal to the sum of the six angles in the two triangles.

The sum of the angle measures in the two triangles = 2 • 180 = 360°.	$a + x + p = 180°$ $c + y + q = 180°$
These six angles make up the four angles of the quadrilateral.	$a + x + p + c + y + q = 360°$ Rearrange terms: $a + (p + q) + c + (x + y) = 360°$
Therefore, the sum of the four angles in the quadrilateral is the same, 360°.	$p + q = b$, and $x + y = d$ Therefore, $a + b + c + d = 360°$.

NUMBER PROPERTIES AND NOTATION

CHAPTER 2

Using Letters in Mathematics

In mathematics, you can use letters to stand for numbers. Letters can have different uses in different situations. Here are four uses of letters:

1. To represent a general number

Example

When you express a rule of arithmetic, such as the commutative property of addition, the letters a and b each stand for any number. $a + b = b + a$

2. To represent a number that you want to find

Example

This equation uses the letter x to represent a number you want to find. $x + 3 = 13$

You want to find a number x that when added to 3 gives 13. In this simple case you see that the number is 10, since 10 is the only number that will make this equation true.

The letter x here is called an *unknown*. If you find a value for x that makes the equation true, you have solved the equation. How could you use this?

Example

Today you have 13 coins in a collection. Someone gave you 3 coins yesterday. How many coins did you have before that? You can represent this problem with the equation $x + 3 = 13$. Solving the equation gives you the answer: $x = 10$ coins.

3. To represent a relationship between two numbers

Example

This equation uses the letters x and y to describe a relationship between numbers. $y = x + 3$

The relationship is that the number y is 3 more than x.

If $x = 1$, then $y = 4$. If $x = 2$, then $y = 5$.

Example

Think about the relationship between two sisters' ages.

If the sisters are three years apart, and the letter x stands for the younger girl's age, and the letter y stands for the older girl's age, you can represent the relationship between their ages with the equation $y = x + 3$. When the younger girl is 1 year old, the older girl is 4 years old. When the younger girl is x years old, the older girl is $x + 3$ years old.

4. To state a general formula (A *formula* is a rule that defines one quantity in terms of other quantities.)

Example

In geometry you learn that the area of a rectangle is equal to its base times its height. This relationship is represented in the general formula $A = b \bullet h$

Here you use the letter A for area, the letter b for the length of one side (called the base), and the letter h for the length of the other side (called the height).

These are some of the ways you use letters in mathematics. There are other ways to use letters that you will learn about as you study algebra.

An *expression* combines arithmetic operations with numbers and/or letters.

Example

These are examples of expressions:

$a + 3b$ $\quad (m^2 + 7)(n^2 + 1)$ $\quad 9.5x - 7.9y$ $\quad pq - 2s^3$

Conventions for Using Letters in Mathematics

A *convention* is an agreed upon way of doing something. Here are some mathematical conventions that are used around the world when using letters.

Multiplication

- When a letter and a number are multiplied, no multiplication sign is needed.

Example

$4n$ means "four multiplied by n" and is read as "four n."

- 1 multiplied by x is written as x. Write x, not $1x$.
- –1 multiplied by d is written as $(-1)d$ or $-d$, but is not written as $-1d$.
- –5 multiplied by c is written as $(-5)c$ or as $-5c$.
- When a letter and a number are multiplied, write the number first and the letter second.

Example

$5 \bullet c$ is written as $5c$, not $c5$.

Powers

- Just as you can write $4 \bullet 4 = 4^2$ and $4 \bullet 4 \bullet 4 = 4^3$, you can write $d \bullet d = d^2$ and $d \bullet d \bullet d = d^3$. The same is true for d^4, d^5, and so on.

- d^2 and d^3 are called powers of d. The small, raised 2 in d^2 and the small, raised 3 in d^3 are called *exponents.*
- You read d^2 as "d squared," d^3 as "d cubed," and other powers, such as d^4, as "d to the fourth power" or just "d to the fourth," and so on.
- An exponent can be a letter too.

Example

d^x ("d to the x power") would represent "d squared" if $x = 2$, "d cubed" if $x = 3$, and so on.

Division

- 2 divided by 3 can be written as $\frac{2}{3}$.
- b divided by c can be written as $\frac{b}{c}$.
- b divided by 5 is written as $\frac{b}{5}$. You can also write $\frac{b}{5}$ as $\frac{1}{5}b$.
- You can write $\frac{3x}{4}$ as $\frac{3}{4}x$.
- $\frac{a}{bc}$ means "a divided by (bc)." $\frac{ac}{b}$ means "ac divided by b."

Parentheses

Parentheses are used to group parts of an expression together when you want them to act as a single quantity. Compute the quantity within the parentheses before performing other operations.

Example

The quantity $(2 \bullet a) \div (5 \bullet b)$ is read as,

"Divide the quantity $(2 \bullet a)$ by the quantity $(5 \bullet b)$."

Example

An expression like $x(y + 4)$ is read as,

"x times the quantity y plus 4."

Remember, the parentheses tell you to treat $(y + 4)$ as a single quantity.

Evaluating an Expression

Suppose pencils cost 20 cents each, and pens cost 30 cents each. If someone buys 8 pencils and 7 pens, the total cost is:

$$(20 \bullet 8) + (30 \bullet 7) = 370 \text{ cents} = \$3.70$$

A general expression can be written to describe the cost of m pencils and n pens.

$$(20m + 30n) \text{ cents}$$

You can use this expression to find the cost for other specific values of m and n.

Example

For example, to find out how much it costs to buy 6 pencils and 4 pens, you can put $m = 6$ and $n = 4$ into this expression. You get:

$$(20 \bullet m) + (30 \bullet n)$$

$$(20 \bullet 6) + (30 \bullet 4) = 240 \text{ cents} = \$2.40$$

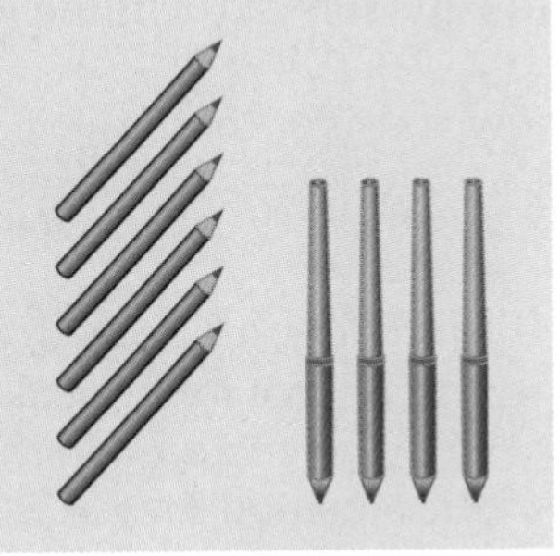

Doing this is called *evaluating* the expression. Specifically, it is evaluating the expression $20m + 30n$ at the values $m = 6$ and $n = 4$.

Terms and Coefficients

The expression $3m + 2y + 4m$ can be used to illustrate what is meant by a "term" and a "coefficient."

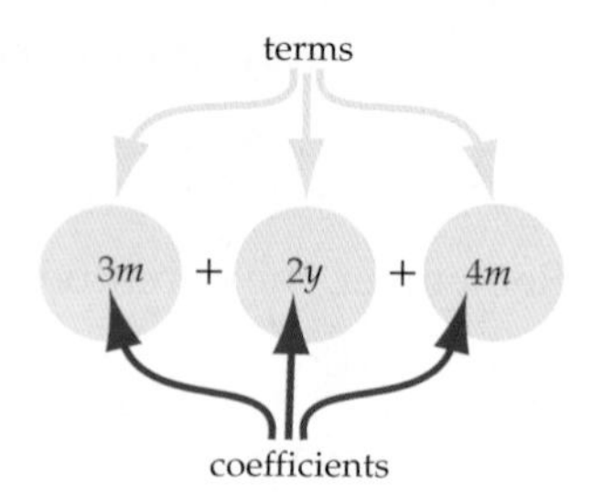

- $3m$, $2y$, and $4m$ are called *terms* of the expression.
- The numerical parts 3, 2, and 4 of the terms $3m$, $2y$, and $4m$ are called *coefficients*.
- m and y are called the *variable* parts of the terms $3m$, $2y$, and $4m$.

The Equal Sign (=)

Consider these two equations:

$$3 + 4 = 2 + 5 \qquad\qquad 7 + 4 = 3 + 6$$

The first equation is true, but the second equation is not true.

An equation is true if the left side is equal to the right side.

Now consider these two equations:

$$3 + x = 2 + 5 \qquad\qquad 7 + 4 = x + 6$$

Until they are solved, neither of these equations is true or false. Until the variable x is given a numerical value, you cannot say whether they are true or false. If you give the value $x = 4$, the equation $3 + x = 2 + 5$ is true. You say $x = 4$ is a *solution* to the equation $3 + x = 2 + 5$.

What is a solution to the equation $7 + 4 = x + 6$?

Number Properties of Addition and Multiplication

	Addition	Multiplication
Identity Property	$a + 0 = a$	$a \bullet 1 = a$
Commutative Property	$a + b = b + a$	$ab = ba$
Associative Property	$a + (b + c) = (a + b) + c$	$a(bc) = (ab)c$
Inverse Property	$a + (-a) = 0$	$a \bullet \frac{1}{a} = 1$
Distributive Property	$a(b + c) = ab + ac$	

Any number multiplied by 0 is 0. The zero property is $a \bullet 0 = 0$.

Identity Property

The number 0 is the *additive identity*; adding 0 to a number does not change the number.

The number 1 is the *multiplicative identity*; multiplying a number by 1 does not change the number.

Inverse Property

For every number a there is a number $-a$, which when added to a gives 0, the *additive identity*.

The number $-a$ is called the *opposite* of a, or the *additive inverse* of a.

Similarly, for every number $a \neq 0$ there is a number $\frac{1}{a}$, which when multiplied by a gives 1, the *multiplicative identity*.

The number $\frac{1}{a}$ is called the *reciprocal* of a, or the *multiplicative inverse* of a.

The inverse properties of addition and multiplication reveal the rule for "undoing" each operation. As is shown in the number properties table, the inverse of adding a number is subtracting that number, and the inverse of multiplying by a number is dividing by that number.

Similarly, the inverse of subtracting a number is adding the number, and the inverse of dividing by a number is multiplying by the number.

Another way to think about dividing by a number a is to think of multiplying by the number $\frac{1}{a}$. $\frac{1}{a}$ is the reciprocal , or multiplicative inverse, of a. Dividing by a is the same as multiplying by the reciprocal $\frac{1}{a}$.

Example

$5 \div 5$ is the same as $5 \bullet \left(\frac{1}{5}\right) = \frac{5}{5} = 1$.

The Distributive Property

The *distributive property* is a rule for using addition and multiplication together. This is an extremely important property that will come up again and again in different contexts.

In general, the distributive property allows you to rewrite the product $a(b + c)$ as the sum $ab + ac$.

You can use the formula for the area of a rectangle to illustrate the distributive property.

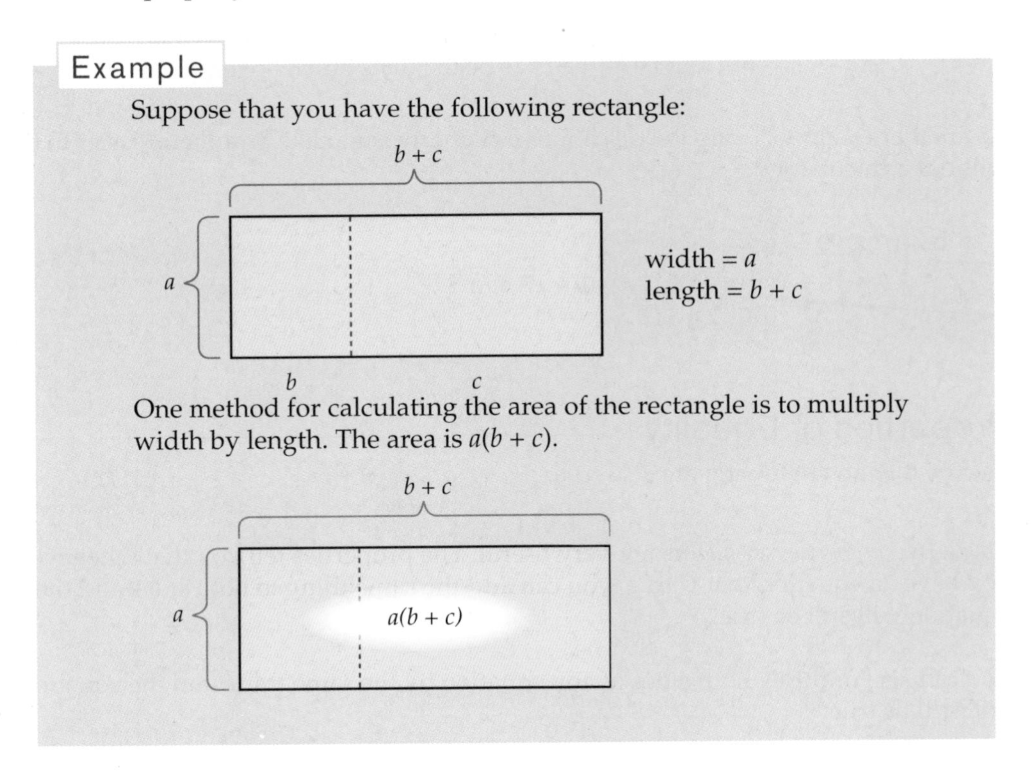

Example

Suppose that you have the following rectangle:

One method for calculating the area of the rectangle is to multiply width by length. The area is $a(b + c)$.

Example

You can also think of this rectangle as being made up of two smaller rectangles. The areas of the two smaller rectangles are ab and ac. The total area is $ab + ac$.

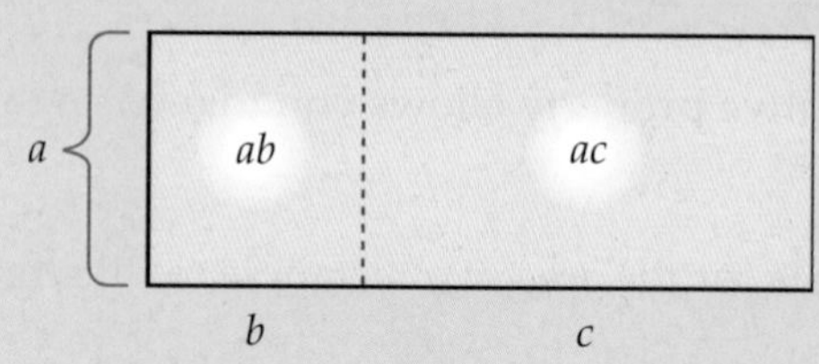

The two results for the area must be the same. This means that $a(b + c) = ab + ac$. The letter a has been "distributed" over $(b + c)$.

As another example, using the distributive property can make arithmetic easier to do without a calculator:

Example

$$7 \bullet 16 = 7(10 + 6) = (7 \bullet 10) + (7 \bullet 6)$$
$$= 70 + 42 = 112$$

Properties of Equality

If $a = b$, then the following are also true:

$$a + c = b + c$$
$$ac = bc$$

These two *properties of equality* are very useful. The properties tell you that whenever you have an equation that is true, you can add the same thing to both sides and the equation will still be true.

You can also multiply both sides of any equation by the same thing and the equation will still be true.

What about subtraction and division?

Subtraction is the same as adding the opposite.

Example

$3 - 2$ is the same as $3 + (-2)$.

Division is the same as multiplying by the inverse.

Example

$3 \div 2$ is the same as $3 \bullet \frac{1}{2}$.

Based on the inverse properties, you can also subtract or divide both sides by the same thing and the equation will still be true.

Example

If $a = b$, then:

$$a - c = b - c$$

$$\frac{a}{c} = \frac{b}{c}$$

Equations

An equation states that the expressions on the left and the right of the = sign express equal values. The two sides represent the same number. Some equations consist of only numbers. Some equations consist of letters and numbers.

An equation can be true or false.

$7 = 5$ is false.
$5 = 5$ is true.

How about $x + 1 = 5$? As it stands it is neither true or false. What value of x will make the equation true? If you put 4 where x is, it is true. Otherwise, it is not. Finding the values that make an equation true is called *solving the equation.*

The Number Properties

Here is a summary of the number properties:

Properties of Addition		**Properties of Multiplication**	
The Identity Property of Addition		*The Identity Property of Multiplication*	
In words	In symbols	In words	In symbols
When you add zero to a number, you get that number.	$a + 0 = a$	When you multiply a number by 1, you get that number.	$a \bullet 1 = a$
The Commutative Property of Addition		*The Commutative Property of Multiplication*	
In words	In symbols	In words	In symbols
The order in which two numbers are added does not affect the sum.	$a + b = b + a$	The order in which two numbers are multiplied does not affect the product.	$ab = ba$
The Associative Property of Addition		*The Associative Property of Multiplication*	
In words	In symbols	In words	In symbols
The sum of three numbers is the same no matter how they are grouped using parentheses.	$(a + b) + c = a + (b + c)$	The product of three numbers is the same no matter how they are grouped using parentheses.	$(a \bullet b) \bullet c = a \bullet (b \bullet c)$

<table>
<tr><th colspan="2">Properties of Addition</th><th colspan="2">Properties of Multiplication</th></tr>
<tr><td colspan="2">The Inverse Property of Addition</td><td colspan="2">The Inverse Property of Multiplication</td></tr>
<tr><td>In words</td><td>In symbols</td><td>In words</td><td>In symbols</td></tr>
<tr><td>Adding a number to its opposite gives zero.</td><td>$a + (-a) = 0$</td><td>Multiplying a non-zero number by its reciprocal gives 1.</td><td>$a \div a = 1$, or
$a \bullet \frac{1}{a} = 1$</td></tr>
<tr><td colspan="4">The Distributive Property</td></tr>
<tr><td colspan="2">In words</td><td colspan="2">In symbols</td></tr>
<tr><td colspan="2">Multiplying a number by the sum of two numbers in parentheses is the same as multiplying the number by each number inside the parentheses and adding the resulting products. It is this property that shows how addition and multiplication interact with each other.</td><td colspan="2">$a(b + c) = ab + ac$</td></tr>
</table>

USING ALGEBRA TO SOLVE PROBLEMS

CHAPTER 3

Use this framework for solving word problems:

- Read the word problem, understand the situation, and understand what questions you are asked. What is the problem situation? Identify the quantities in the situation. Name and label those quantities.
- Represent the problem situation using some of these tools:

  Make diagrams with labels or pictures of relationships.

  Describe the problem situation in your own words.

  Make tables.

  Try out simple numbers.

  Break the problems into smaller parts.

  Make up equations to express relationships.

- Explain your representations to other students; listen to other students' explanations and understand their representations.
- Answer questions about the problem. This may include developing equations to answer the questions posed, making any needed calculations, and finding the answer to any question asked in the problem.

Understand and represent the problem situation. Do not race toward an answer. The purpose of problem solving is to learn mathematics and to get better at solving new and unfamiliar problems.

Understanding the Problem Situation

A word problem describes a situation. First, understand the situation described. Imagine the situation described in the problem, just like you would imagine the situation when you read a story. In a story, you focus on the characters, their motives, and the actions and reactions of characters. The action in a math problem, however, is not between characters. The action in word problems is between quantities.

To understand a story, a reader has to recognize which characters and events are being referred to.

Example

Someone took his jacket. His Dad will be mad. But it wasn't his fault. He left it on his chair.

He saw his best friend, Johnny, and ran up to him.

"Someone took my jacket. Let's find him and…"

"You just did."

Johnny reached into his backpack…

You can imagine the situation caused by someone taking a jacket.
Can you answer these questions about the situation?

1. Who are the people referred to in this episode? Assign a name to each.

   Someone who took jacket = mystery person (or "unknown")
   The boy whose jacket was taken = boy
   The boy's friend Johnny = Johnny
   The boy's Dad = Dad
   There are four people in this little story. But there is a surprise!

2. Who took the boy's jacket?

   When the boy says "Let's find him…," *him* refers to the mystery person. Then Johnny says, "You just did." What does Johnny mean? "Did" what? He means "You just found him, the mystery person." But he just found Johnny! Therefore:

   mystery person = Johnny

Johnny took the jacket. So the mystery person and Johnny turn out to be the same person. But when you read the story, you do not know this until Johnny tells you. This means there are only three people in the story.

Word problems work the same way as stories. However, instead of describing characters, they describe quantities. When you read a word problem, you have to understand and keep track of what quantities are being described.

What is a quantity? A *quantity* is something that can be counted or measured. A quantity has a description of what is being measured or counted (inches, height in feet, weight in pounds, apples, dollars) and a number (or letter representing an unknown or variable).

Example

"John's height" is a quantity. A number and unit can be assigned to this quantity.

"John's height is 64 inches." This means that "John's height = 64 inches." "John's height," and "64 inches" are the same; they describe the same quantity.

Suppose you are also told that John's height is 10 inches more than Sonia's. If you already understand that 64 inches and John's height describe the same quantity, then you can figure out that 64 inches is 10 more inches than Sonia's height.
Sonia's height = John's height – 10 inches = (64 – 10) inches = 54 inches.

Here is a situation:

Example

Mrs. Jackson has 27 stickers. She needs 7 more in order to give 2 stickers to each student in her class.

This situation describes:

1. the number of stickers that Mrs. Jackson has;
2. the number of stickers she needs;
3. the number of stickers each student gets.

And, even though it does not say so directly, the situation is also about the number of students in Mrs. Jackson's class. This is a problem situation about quantities and how they relate to each other.

How can you imagine and represent this situation?

> Imagine: a pile of 27 stickers, giving 2 to each student, needing 7 more so every student has 2. A diagram can help.

Representing the Situation

Diagrams

A diagram of a situation should show and label the relevant quantities.

In the sticker problem, one way to make a diagram is to pair the stickers to show the "2 for each student." This leaves the odd or unpaired 27th sticker as a single. The "7 more" shows 3 pairs and a single that can be matched off with the single left over from the original 27 stickers.

Each pair represents 2 stickers and 1 student's share.
The diagram shows that 2 • (number of students) = number of stickers.

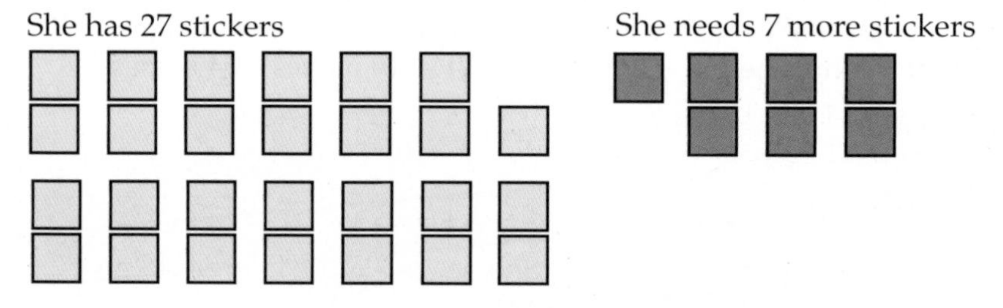

2 stickers = 1 student's share

Tables

Another good way to show a relationship between quantities is by using a table. Tables are excellent tools for situations involving two quantities that change in relation to each other.

Think about the relationship between the number of students and the number of stickers if the number of students could vary. Thinking this way helps you see the relationship and gets you ready to solve other problems you have not seen yet.

Example

Show how many stickers are needed for 1 to 10 students if each student gets 2 stickers.

A table shows this clearly:

No. of Students	1	2	3	4	5	6	7	8	9	10	n
No. of Stickers	2	4	6	8	10	12	14	16	18	20	$2n$

Language

Language is another helpful representation.

In most problems, the same quantity is expressed in two or more ways, such as a "mystery person" and "Johnny" or "Sonia's height" and "John's height – 10." Knowing how the same quantity is described in different ways will help you understand the situation.

Example

Both lines below refer to the same quantity:

(stickers she has) + (stickers she needs) = 2 • (number of students)

The very important relationship between the quantities is:

total stickers needed = twice the number of students

Another relationship is:

the total number of stickers = the sum of how many she has and how many she needs

Another relationship is a simple rate:

two stickers per student

Equations

Mathematics adds some tools to language that make it easier to use and understand the language. Those tools include equations and letters.

Example

"Mrs. Jackson has 27 stickers. She needs 7 more in order to give 2 stickers to each student." This statement can be reorganized as follows: (stickers she has) + (stickers she needs) = 2(number of students). Written as an equation where n = the number of students:

$$27 + 7 = 2n$$

The equation expresses the central relationship among the quantities. It helps you understand what is really going on in this situation. The diagram worked for the specific numbers in the sticker problem and showed the pairing well. The table worked to show how the number of stickers related to the number of students.

To show the relationship for any number of students, the equation tells the story best:

$$S = 2n$$

where S is the number of stickers, and n is the number of students.

Graphs

An important part of learning to represent a variety of situations is understanding how different representations of the same situation correspond. In the sticker situation, you need to understand that the "2" in "$2n$" corresponds to the "2-ness" of the pairs in the diagram, and that the "7" corresponds to the 7 more stickers needed.

Graphing is another essential tool for representing situations. A graph shows the relationship between two quantities that vary for all the values of the variable within the domain and range of the graph. Graphs are fundamental representations for functions and equations. Making sense of graphs builds on a solid understanding of the number line.

This graph represents the equation $S = 2n$.

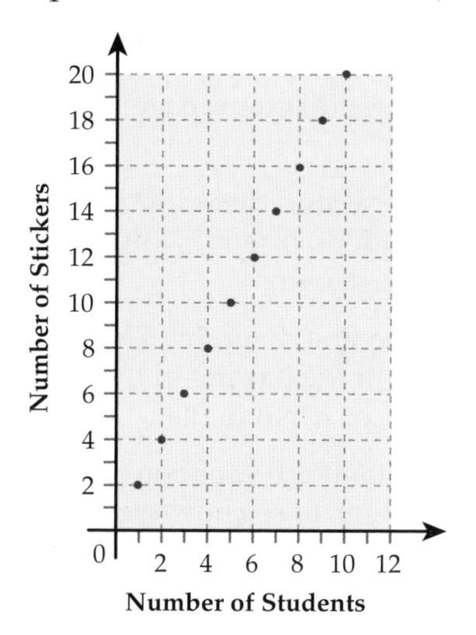

Understanding Representations of Others

Different students will think about problem situations in different ways. Practice explaining your representations of problem situations to your partner and the class. Try to understand how other students think about, and represent, the same situation.

By understanding other students' ways of thinking, you will learn different ways to represent and understand the problem. Each representation is like a window through which you see the same situation from a different perspective.

Students will create different diagrams. One student might make a diagram something like the sticker diagram on page 26, where each sticker is represented. Another student might make a diagram that shows each student, instead of each sticker.

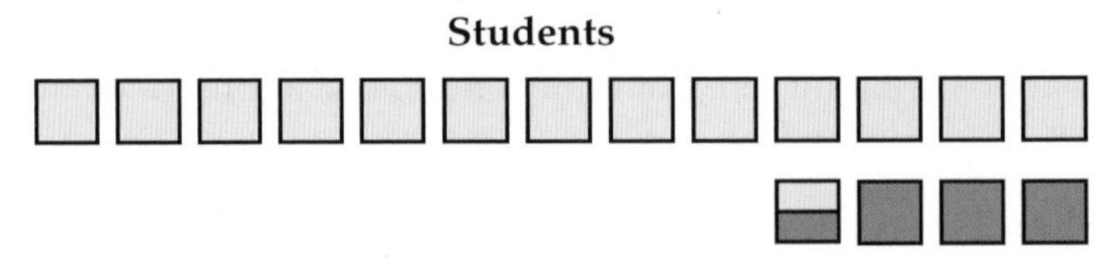

How do these diagrams correspond?

- Each pair of stickers in the first diagram (page 26) corresponds to a student in the drawing above.
- Where are the stickers in the second diagram? You could write a "2" on the square representing each student, or put 2 little marks on each to show how the stickers relate to the students shown.
- Where are the students in the first diagram? The pairing represents a student's share. Look for the "2-ness" of the pairing in the equation. "$2n$" means 2 stickers for each student times "n" students.

Another important way to understand the situation is to examine models of the situation developed by other students to see how their models work.

- Try out different values for the quantities and see what happens.
- Try very small values (less than 1); try 1, 2, 10; try a large number like 1,000. See what happens.
- Ask students which values make more sense, given the situation.
- Try different kinds of numbers if possible: fractions, decimal numbers, and whole numbers. For example, in the sticker situation, a table is a good way of writing down different values you have tried.

Answering Questions about the Problem Situation

Equations are often the easiest way to represent a situation that you already understand. But they are abstract and difficult to construct until you understand the situation. Once you realize that two different expressions for quantities in the situation are referring to the same quantity, you can show that one expression "equals" the other. That is an equation.

Once you have imagined the situation, many questions will make sense:

How many stickers are needed, in total, so that every student has 2?

> The diagram on page 30 shows the number of students.
> Count the number of students and multiply by two.
>
> Two times the number of students is the same number as the number of stickers. If you know the number of students, n, the formula $S = 2n$ will work for any number of students.

How many students are there in the class?

> Counting the pairs of stickers tells you how many students are in the class.
>
> Half the number of stickers is the same number as the number of students. You can divide the total number of stickers by 2.

If three more students joined the group, how many more stickers would be needed?

> Three students need three more pairs. Add $2 \bullet 3 = 6$ stickers to the previous total. The formula $S = 2n$ will work for any number.

What if Mrs. Jackson wanted to give 5 stickers to each student? How many stickers would she need?

> 5 stickers per student could be represented by changing the pairs in the diagram on page 26 to groups of 5.
>
> The number of stickers would be 5 times the number of students. The formula would change from $S = 2n$ to $S = 5n$.

Understanding and Solving Common Problem Types —

Problem Situations with a "Before→Action→After" Structure

Example

Dwayne and Lisa are using carts to move floor tiles.

Each tile weighs $\frac{1}{2}$ pound.

Lisa's cart already has a load of $62\frac{1}{2}$ pounds.

Dwayne's cart only has $11\frac{1}{2}$ pounds.

1. How many tiles have to be moved from Lisa's to Dwayne's cart so their loads have the same weight?
2. How much will each load weigh when they are the same?

This situation has a "before → action → after" pattern, like many stories. The situation described above gives quantities and values for the "before" part of the story. Your approach to resolving Lisa's problem will be the action: you will move tiles. Your action will change the weight on the carts. The new weights on the carts will be quantities in the "after" situation.

You can avoid some confusion with "before → action → after" problems by thinking about which quantities will change their values from before to after and which quantities will not change.

Which quantities will change from before to after in this situation?
These are the quantities in the before situation:

Lisa's load in pounds	l	$= 62\frac{1}{2}$ lb
Dwayne's load in pounds	d	$= 11\frac{1}{2}$ lb

Which quantities will not change?

You do not need to assign letters to quantities that you know will not change (but you can, if it helps you think about it).

Weight of a tile in pounds $t = \frac{1}{2}$ lb

These unchanging quantities are not mentioned but could be important:

Number of tiles per pound $= \frac{(1 \text{ lb})}{t} = \frac{(1 \text{ lb})}{\frac{1}{2} \text{ lb per tile}} = 2$ tiles

The sum of Lisa's load and Dwayne's load $= l + d = 62\frac{1}{2} + 11\frac{1}{2} = 74$ lb.

How could you make a diagram that shows these quantities and how they relate to each other?

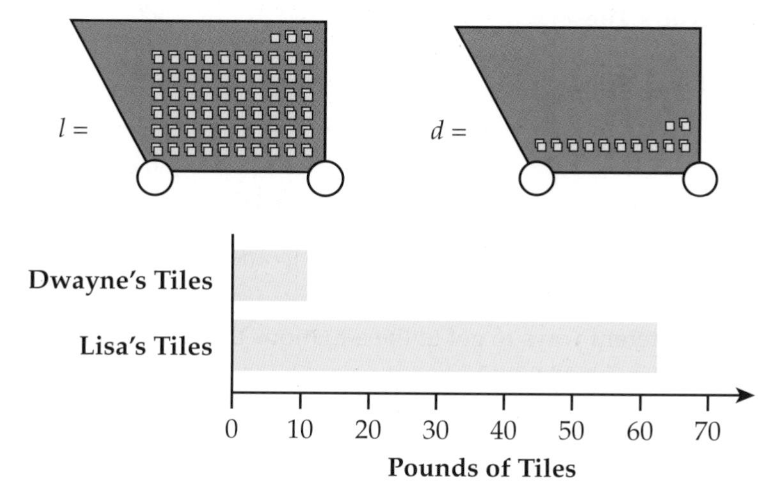

The relationships between quantities can be shown in various ways: by relative size, by connecting lines, by circling corresponding quantities, and by arrows.

Here is a before/after diagram.

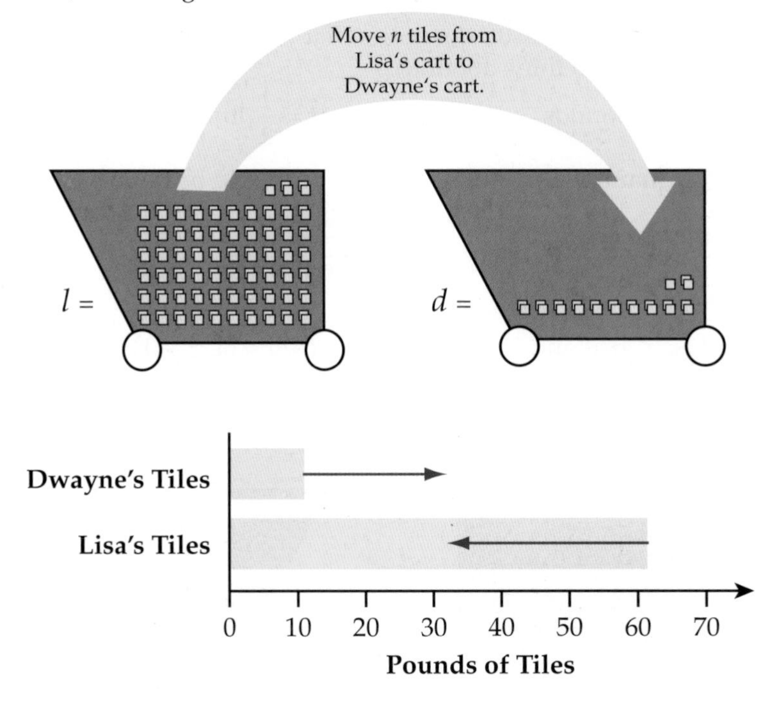

How could you show the relationships that will help answer the questions using equations?

There are several different ways to get to the equations that will answer the questions in the problem. Here is one way.

Look at the questions. What quantities does the problem ask for?
Are there any quantities you have not already named?
Yes, there are two:

The number of tiles moved	$= n$
The weight of each load when the loads are equal	$= q$

The next quantity is not mentioned, but could be important, because so much information is given about the weights.

Weight of the tiles moved $= b$

To write equations, look for two different ways of expressing the same quantity.

Example

The total weight of both original loads can be expressed as $l + d$. After you move the tiles, each load's weight $= q$, so the total weight can be expressed as $q + q$. Since the total weight of the two loads does not change when you move them,

$$q + q = l + d \text{ or } 2q = l + d$$

There is another way of expressing the total weight of the loads:

$$l + d = 62\frac{1}{2} + 11\frac{1}{2} = 74 \text{ lb}$$

This total does not change from before to after.

$$2q = 74 \text{ lb}$$
$$q = 37 \text{ lb}$$

These equations answer the question: How much will each load weigh when the two loads are the same?

Each load will weigh 37 pounds when the two loads are the same.

Question 1 asks for the number of tiles. You have been working with pounds. You have already calculated a conversion factor of 2 tiles per pound. (See Chapter 16, *Units and Quantities*.)

$$2\,\frac{\text{tiles}}{\text{lb}}$$

You can use this conversion factor to write an equation for n, the number of tiles moved, in terms of b, the weight of tiles moved:

$$n = 2b \qquad \text{with units: } n \text{ tiles} = 2\,\frac{\text{tiles}}{\text{lb}} \bullet b[\text{lb}]$$

You know Lisa's load before, $62\frac{1}{2}$ lb, and after, 37 lb. The difference is how much has to be moved, b.

$$b = 62\frac{1}{2} - 37 = 25\frac{1}{2} \text{ lb}$$

Substitute to find n:

$n = 2 \bullet 25\frac{1}{2}$ with units, n tiles $= 2\,\frac{\text{tiles}}{\text{lb}} \bullet 25\frac{1}{2}$ lb

$n = 51$ tiles $\frac{\text{lb}}{\text{lb}}$ cancels out, leaving tiles

This answers question 1: 51 tiles have to be moved from Lisa's cart to Dwayne's cart to make their loads equal.

There are other ways to arrive at the same answer.

Example

You could convert the pounds of weight to the number of tiles before doing any calculations.

$l = 62\frac{1}{2}$ lb = 125 tiles

$d = 11\frac{1}{2}$ lb = 23 tiles

$l + d = 74$ lb = 148 tiles

$\frac{1}{2}(l+d) = 37$ lb = 74 tiles

The number of tiles that need to be moved to make the carts equal is:

74 tiles = 125 – 74 = 51 tiles

You can summarize the relationships, including the answers to the questions, in a before ➔ action ➔ after table:

Quantity	Before	Action	After
Lisa's load	$l = 62\frac{1}{2}$ lb	remove $25\frac{1}{2}$ lb of tiles	$q = 37$ lb
Dwayne's load	$d = 11\frac{1}{2}$ lb	add $25\frac{1}{2}$ ls of tiles	$q = 37$ lb
Total load	$l + d = 74$ lb	none	$l + d = 74$ lb
Total split equally, q	$\frac{(l + d)}{2} = 37$	none	$\frac{(l + d)}{2} = 37$ lb
Lb of tiles moved	none	move $25\frac{1}{2}$ lb of tiles	$25\frac{1}{2}$ lb
Tiles per lb	2	none	2
Lb per tile	$\frac{1}{2}$	none	$\frac{1}{2}$
Number of tiles in $25\frac{1}{2}$ lb	51	none	51
Number of tiles moved, n	none	move tiles	51

Problem Situations with a Base Value and a Rate

Example

1. Sheila bought $40 worth of skirts at the weekend sale. When she brought her skirts to the counter, the clerk added an 8% sales tax. How much was Sheila's bill?
2. Lisa bought a new pair of shoes. She paid $43.20, which included an 8% sales tax. What was the price of the pair of shoes before sales tax?

These problems are a special kind of "before→action→after" problem. The action is a two-step action:

1. Multiply the before quantity times a rate (in this case, the rate is the sales tax, 8% or 0.08 times the price); and then
2. Add the product to the before quantity to get the after quantity.

It is easier to describe and see the situation like this:

$$(\text{before} \bullet \text{rate}) + \text{before} = \text{after}$$

Notice how the distributive property can be applied to this equation to show it in a different form that can make calculation easier. Factor out the "before," and the equation becomes:

$$\text{before}(1 + \text{rate}) = \text{after}$$

In this example:

$$(\text{price} \bullet \text{tax}) + \text{price} = \text{cost}$$
$$0.08p + p = c$$

Using the distributive property:

$$\text{price}(1 + \text{tax}) = \text{cost}$$
$$p \bullet 1.08 = \text{cost}$$
$$1.08p = c$$

These two formulas are equivalent; they give the same answer every time, guaranteed by the distributive property, but the second is easier because adding 1 is not as difficult as adding whatever p is.

Example

Problem 1 gives the price and the tax rate:

$$\$40 + (0.08 \bullet \$40) = \$40 + \$3.20 = \$43.20 = \text{Sheila's bill}$$

$$1.08 \bullet \$40 = \$43.20 = \text{Sheila's bill}$$

Example

Problem 2 gives the after (the bill) and the rate and asks you to recover the before (the price). Starting with the same equation, you want to get the unknown by itself, with all the knowns on the other side of the = sign.

$c = rp + p$ or $c = p(1 + r)$

$43.20 = 1.08p$ divide both sides by 1.08 to get p by itself

$43.20/1.08 = p$

$\$40 = p$

Here are some other kinds of problem situations that have this same kind of "before (1 + rate) = after" structure:

Situation	Basic Structure	Constant Rate	Equations	Example Units
Charges	taxes, interest, fees based on % new cost = (tax rate • old cost) + old cost	tax rate	$c = p + (r \bullet p)$	$ = $ + (% • $)
	new price = old price – (discount rate • old price)	discount rate	$c = p - (r \bullet p)$	Note: % is [$/$], which is unitless.
Enlargements and reductions	new size = old size • (1 + % enlargement)	% enlargement	$s = q(1 + r)$	cm = cm
	new size = old size • (1 – % reduction)	% reduction	$s = q(1 - r)$	Note: $(1 \pm r)$ is unitless.

Problem Situations with Continuing Action

Suppose the situation with the tiles and carts from the "before → action → after" structure had this feature added:

Lisa and Dwayne can move 5 tiles a minute.

In the earlier version, no information about time to move the tiles was given. When this information is added, new kinds of questions can be asked, and new information can be represented:

1. How long did it take to make the loads equal?
2. How many tiles have been moved after t seconds?

Question 2 describes the relationship between elapsed time and the number of tiles moved. This relationship is expressed as a rate, "5 tiles per minute." This rate itself is a third quantity. The units of this rate are "tiles/minute." In this problem situation, the rate does not change.

A relationship that has a constant rate is often represented with an equation, a table, and a graph. When the number of minutes can vary, and the rate of tiles per minute is constant (invariant), then the number of tiles will vary in a predictable way:

$$n = 5t$$

where t is the amount of minutes, and n is the number of tiles moved.

Many constant rate problem situations have this kind of simple structure:

$$n = rt$$

where n is a quantity, r is a rate quantity, and t is another quantity, like time.

The rate, r, is a ratio of $\frac{n}{t}$. It expresses the constant relationship between n and t.

From this simple structure, many problems can be posed. As a problem solver, it helps to think through the situation so you can see and express the structure.

Example

The height, h, of a stack of books (or nickels, or boards, or paper) = the thickness, t, of each book times the number, n, of books.

$$h = tn$$

In this situation the thickness is the rate "inches per book."

From this situation you can get three simple problems:

1. What is the height given thickness and number? $h = tn$

Example

What is the height of a stack of 32 books $\frac{3}{4}$ inch thick?

$$h = \frac{3}{4} \text{ of an inch per book} \bullet 32 \text{ books}$$

2. What is the thickness given height and number? $t = h/n$

Example

How thick is each book in a stack of 32 books that is 24 inches high?

$$t = \frac{32}{24} \text{ inches}$$

3. What is the number given height and thickness? $n = h/t$

Example

How many books are in a stack 24 inches high if each book is $\frac{3}{4}$ inch thick?

$$n = 24 / \frac{3}{4} \text{ books}$$

You end up with three equations. You start with the equation $h = tn$ which expresses a multiplication (of t times n). There are three quantities, and one is the product of the other two. Every multiplication leads to two possible divisions, $h/t = n$ and $h/n = t$.

Below is a table that shows common types of problems like the one discussed above. These relationships are proportional (See Chapter 19, *Proportional Relationships*) because the value of one variable can be calculated by the value of the other variable times a constant rate. The relationships can be expressed by equations that look like this, $y = kx$, where k is the constant rate.

Examples of Proportional Relationships

Situation	Basic Structure	Constant Rate	Equations	Example Units
Stacks	height = thickness • number of items	thickness per book	$h = tn$ $t = h/n$ $n = h/t$	inches = (inches/item) • # items
Groupings	total number = (# per group) • (# of groups) e.g. legs = (legs per chair) • # of chairs	number per group	$t = gn$ $g= t/n$ $n = t/g$	eggs = (eggs/carton) • # cartons
Scales (on a map or drawing)	# of miles = miles per inch • # of inches	miles per inch	$M = (m/i) \bullet I$ $M/(m/i) = I$ $M/I = m/i$	miles = (miles/inch) • inches

Prices	cost = price • # of items
	cost = price • # of lb
	$ per item
	$ per lb
		$ per item
		$ per lb
			$c = pn$
			$p = c/n$
			$n = c/p$
				$ = ($/item) • # items
				$ = ($/lb) • # lb

Speeds	distance = speed • time	miles per hour	$d = rt$ $r = d/t$ $t = d/r$	miles = (miles/hour) • hours
Similar figures	base of triangle *A* = ratio of similarity • base of triangle *B*	length *A* per length *B*	$A = (a/b)B$ $a/b = A/B$ $B = A/(a/b)$	inches = (inches/inches) • inches
Slopes (that start at 0 height and 0 length)	rise = (rise/run) • run run = (run/rise) • rise	height of rise per length of run	$y = mx$ $m = y/x$ $x = y/m$	cm = (cm/cm) • cm or unitless

Problem Situations Where the Rate Is Not Explicit

Example

Ann bought 4 lb of onions for $3.56 At the same price as onions, how much would 10 lb of potatoes cost?

The price is the invariant in this situation: it does not change. But the price is not mentioned in the problem. It is implied.

The cost changes when the number of pounds change; the relationship between cost and the number of pounds does not change. That relationship is called the price:

p = price; c = cost; n = number of pounds

$$p = c/n$$

A good way to think about problems like this one is to recognize the invariant. If it is implied, figure out what it is:

price = $3.56/4 lb	given by the problem, cost per 4 lb
price = $ 0.89/lb	divide $3.56/4 to get cost per lb

Once you know the price, you can find the cost for any number of pounds.

$$c = pn$$

$$c = \$.89 \bullet 10$$

c = $8.90 for 10 lb of potatoes

Here is another example of an implied invariant that is a constant rate:

Example

It took 20 seconds to fill 3 gallons from a hose.
How long would it take to fill 50 gallons using that hose?

This problem involves time, like many rate problems do. What is the invariant quantity? It is the rate at which gallons are filled, gallons per second.

$r = n$ gallons/second

This rate is not mentioned, but it is implied in the information,
The rate is 3 gallons per 20 seconds.

$$r = n \text{ gal}/t \text{ sec} = 3 \text{ gal}/20 \text{ sec} = 0.15 \text{ gal}/\text{sec}$$

At that rate, how long will it take to get 50 gallons?
You know that the number of gallons filled = rate • time:

$$g = rt$$

But the question is not asking for gallons. It gives the gallons. It asks for the time.
Divide both sides by r to get:

$$t = g/r$$

Remember that the relationship expressed by the multiplication equation $g = rt$ can also be expressed using either of two division equations, $t = g/r$ and $r = g/t$.
In this problem you want to find time, t, so you want the equation that tells you what t equals:

$$t = g/r$$

$t = 50 \text{ gal}/(.15 \text{ gal}/\text{sec})$	
$t = (50/.15)\ [\text{gal}/(\text{gal}/\text{sec})]$	Collect the units to the right of the numbers
$t = (50/.15) \text{ sec}$	Cancel out gal/gal
$t = 333 \text{ seconds}$	It takes 333 seconds, or 5 minutes and 33 seconds, to fill 50 gallons.

PLACE VALUE

CHAPTER 4

Place Value of Whole Numbers

The number system you use is called the Hindu-Arabic system of numeration.

The ten symbols below can be combined to represent any number in the Hindu-Arabic system. Each of the symbols is called a *digit*.

0	1	2	3	4	5	6	7	8	9

The set of *whole numbers* is {0, 1, 2, 3, 4, 5, 6, 7, 8, 9, 10, 11, 12,…}
Zero is the first whole number. There is no largest whole number; given any whole number, there is always a number that is larger.

Example

6 is a single-digit whole number.
435 and 10,566,342 are multidigit whole numbers.

The numbers are read from left to right.

Example

435 is read as "four hundred thirty-five."
10,566,342 is read as "ten million, five hundred sixty-six thousand, three hundred forty-two."

What Is Place Value?

In the Hindu-Arabic number system, the position (or place) of a digit within a number determines the value of the digit. The term *place value* refers to the importance of the position of a digit in a number.

Moving from right to left, the value of every place in a number increases by a factor of ten.

Moving from right to left, the names of the places are *ones, tens, hundreds, thousands, ten thousands,* and so on. These are *powers of ten*:

Ten thousands Place	Thousands Place	Hundreds Place	Tens Place	Ones Place
10,000 = 1,000 • 10	1,000 = 100 • 10	100 = 10 • 10	10 = 1 • 10	1

Example

Think about the number 1,027. What is the value of each of the four digits?

Here is the number 1,027 displayed in a place-value table:

Thousands	Hundreds	Tens	Ones
1	0	2	7

- The digit 7 is in the ones place. This means there are 7 ones. Its value is 7.
- The digit 2 is in the tens place. This means there are 2 tens. Its value is 20.
- The digit 0 is in the hundreds place. This means there are no hundreds. Its value is 0.
- The digit 1 is in the thousands place. This means there is 1 thousand. Its value is 1,000.

In words, 1,027 is "one thousand, twenty-seven."

You can also express this number as a sum of the values of each of the four places:

$$1{,}027 = 1{,}000 + 0 + 20 + 7$$

Rounding

The numbers 10, 100, 1,000, and so on, are *powers of ten*.

A number that has a value that lies between powers of ten is sometimes rounded up or down to the nearest power of ten. This is done when estimating, when knowing the exact value of the number is not necessary.

Rounding Numbers with a Number Line

You can round any whole number to the nearest ten, hundred, thousand, and so on, if you understand the powers of ten and their multiples.

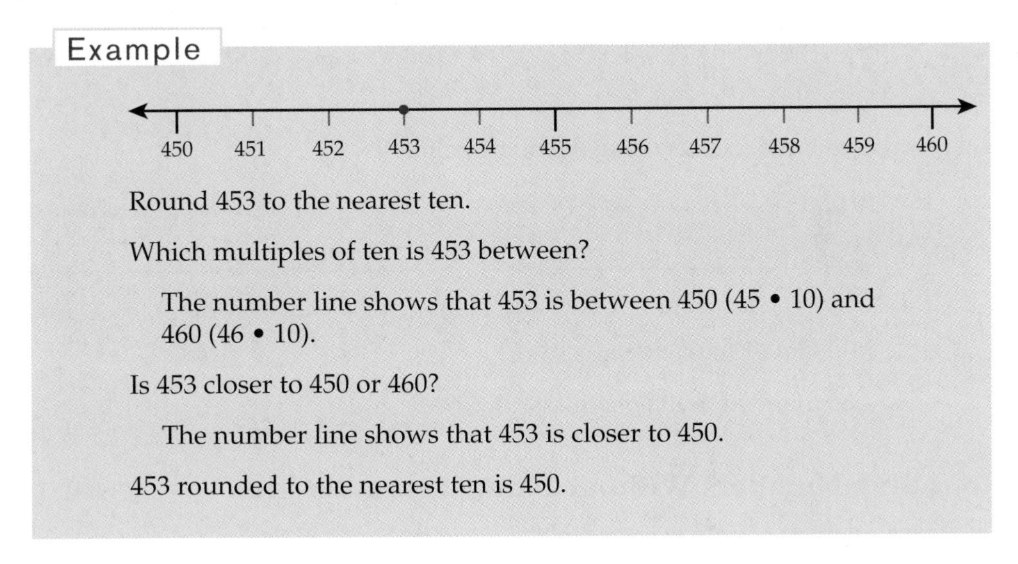

Round 453 to the nearest ten.

Which multiples of ten is 453 between?

The number line shows that 453 is between 450 (45 • 10) and 460 (46 • 10).

Is 453 closer to 450 or 460?

The number line shows that 453 is closer to 450.

453 rounded to the nearest ten is 450.

Example

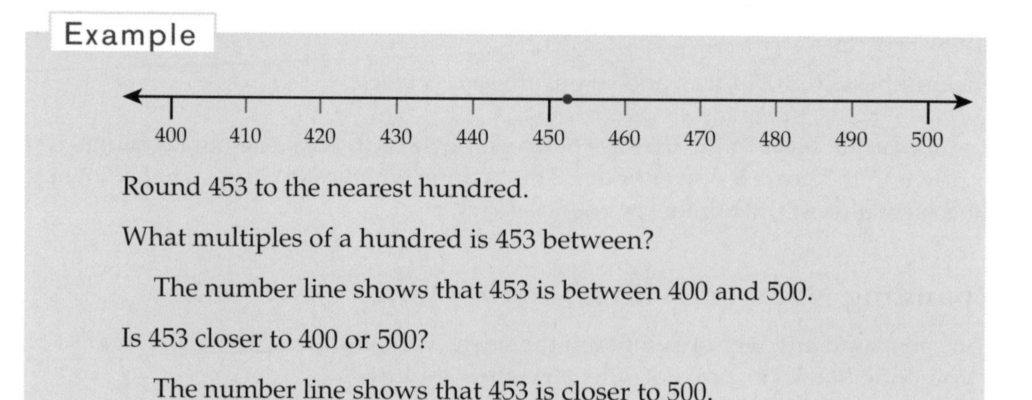

Round 453 to the nearest hundred.

What multiples of a hundred is 453 between?

The number line shows that 453 is between 400 and 500.

Is 453 closer to 400 or 500?

The number line shows that 453 is closer to 500.

453 rounded to the nearest hundred is 500.

You can also round decimals using the number line.

Example

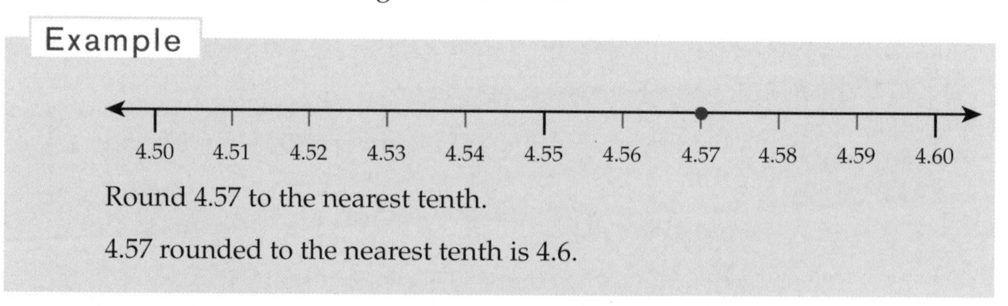

Round 4.57 to the nearest tenth.

4.57 rounded to the nearest tenth is 4.6.

Rounding Numbers Without Using a Number Line

To *round* a number:

- Look at the digit in the place that is one place to the right of the place to which you are rounding.

  If you are rounding to the nearest ten, then look at the digit in the ones place.

If you are rounding to the nearest hundred, then look at the digit in the tens place, and so on.

- If the digit in the place to the right is 0, 1, 2, 3, or 4, then round the number *down*.
- If the digit in the place to the right is 5, 6, 7, 8, or 9, then round the number *up*.

Rounding numbers is a useful method of estimating answers to mathematics problems. Multiples of 10, 100, or 1,000 are easy to work with. If you estimate, or approximate, a number to a multiple of 10, calculating often becomes easier.

Example

Round 6,835

To the Nearest Thousand	To the Nearest Nundred	To the Nearest Ten
6,835	6,835	6,835
Look at the digit 8. Round up to the nearest multiple of a thousand.	Look at the digit 3. Round down to the nearest multiple of a hundred.	Look at the digit 5. Round up to the nearest multiple of a ten.
Round 6,835 to 7,000.	Round 6,835 to 6,800.	Round 6,835 to 6,840.

Expanded Form

When a number is expressed in *expanded form,* it is represented as a sum of products of single-digit numbers multiplied by powers of ten.

Writing a number in expanded form allows you to see the place value of that number's digits.

Comment

Expanded form is also called *expanded notation.*

Example

The number 1,027 is one thousand, twenty-seven. 1,027 expressed in expanded form is:

$$1{,}027 = (1 \bullet 1{,}000) + (0 \bullet 100) + (2 \bullet 10) + (7 \bullet 1)$$

Expanded form can also be written as:

$$1{,}027 = 1{,}000 + 0 + 20 + 7$$

Numbers that are in expanded form can be compressed into standard Hindu-Arabic form.

Example

A number in expanded form is:

$$(6 \bullet 10{,}000) + (0 \bullet 1{,}000) + (9 \bullet 100) + (3 \bullet 10) + (1 \bullet 1)$$

The number in standard form is 60,931.

$(6 \bullet 10{,}000) + (0 \bullet 1{,}000) + (9 \bullet 100) + (3 \bullet 10) + (1 \bullet 1) = 60{,}931$

In words, this number is "sixty thousand, nine hundred thirty-one."

Example

Here are three ways to represent a number.

Words	Sixty-thousand, nine hundred thirty-one
Standard Form	60,931
Expanded Form	(6 • 10,000) + (0 • 1,000) + (9 • 100) + (3 • 10) + (1 • 1) or 60,000 + 0 + 900 + 30 + 1

Ordering Numbers

Place value allows you to compare the values of numbers and put them in order. The symbol ">" means *is greater than*, and "<" means *is less than*.

Symbols	Meaning
>	is greater than
<	is less than

Example

The numbers 7,120 and 1,027 have the same digits, but the two numbers do not have the same value. Using expanded form, you can see why 7,120 is much greater in value than 1,027.

$7{,}210 = (7 \bullet 1{,}000) + (2 \bullet 100) + (1 \bullet 10) + (0 \bullet 1)$

$1{,}027 = (1 \bullet 1{,}000) + (0 \bullet 100) + (2 \bullet 10) + (7 \bullet 1)$

7,210 is the larger number because there are 7 thousands, compared to 1 thousand in 1,027. In this example, to determine the greater number, it is not necessary to consider the hundreds, tens, or ones places.

$$7{,}210 > 1{,}027 \quad \text{or} \quad 1{,}027 < 7{,}210$$

Example

The number 354 is greater than the number 285, because there are 3 hundreds in 354 and only 2 hundreds in 285. To determine the greater number, there is no need to consider the tens or ones places.

Expanded form helps you see this.

$354 = (3 \bullet 100) + (5 \bullet 10) + (4 \bullet 1)$

$285 = (2 \bullet 100) + (8 \bullet 10) + (5 \bullet 1)$

$$354 > 285 \quad \text{or} \quad 285 < 354$$

Example

Compare 3,546 to 3,846. Both numbers have 3 thousands. So you need to compare the digits in the hundreds place. You can use the expanded form of each number to compare the digits in the hundreds place.

$$3{,}846 = (3 \bullet 1{,}000) + (8 \bullet 100) + (4 \bullet 10) + (6 \bullet 1)$$

$$3{,}546 = (3 \bullet 1{,}000) + (5 \bullet 100) + (4 \bullet 10) + (6 \bullet 1)$$

There are 5 hundreds in 3,546, compared to 8 hundreds in 3,846.

$$3{,}546 < 3{,}846 \quad \text{or} \quad 3{,}846 > 3{,}546$$

Place Value of Decimals

Decimals, like fractions, show you parts of whole numbers.

Digits in decimal numbers have values that are determined by the position (or place) of the digit in a number, just like the digits in whole numbers.

Example

48.675

The digits 4 and 8 form the whole number part: 48
The digits 6, 7, and 5 are the decimal part: 0.675.

You use a decimal point to distinguish the whole number part of a number from the decimal part. Any digits that are to the left of the decimal point are in whole number places. Any digits that are to the right of the decimal point are in decimal places.

Example

In the number 48.675, the digit 6 is in the first decimal place, the digit 7 is in the second decimal place, and the digit 5 is in the third decimal place.

In whole numbers, the value of each place is ten times larger than the value of the place to its right. This is also true in the decimal places of numbers. Decimal place values are based on decreasing powers of ten as you go to the right.

Tenths	$1 \div 10 = \frac{1}{10} = 0.1$
Hundredths	$\frac{1}{10} \div 10 = \frac{1}{100} = 0.01$
Thousandths	$\frac{1}{100} \div 10 = \frac{1}{1000} = 0.001$

There is no end to the possible number of decimal places. It is possible to use an infinite number of decimal places to represent a number.

You can write decimals in words.

Example

In words, 3.5 is "three and five-tenths."

In words, 18.34 is "eighteen and thirty-four hundredths."

In words, 48.675 is "forty-eight and six hundred seventy-five thousandths."

Comment

The "and" in each statement represents the decimal point.

Comment

You may hear people say "forty-eight point six seven five" because it is easier, but "forty-eight and six hundred seventy-five thousandths" tells you the place value.

You can write decimals in a place-value table:

Decimal	Tens	Ones	.	Tenths	Hundredths	Thousandths
3.5		3	.	5		
18.34	1	8	.	3	4	
48.675	4	8	.	6	7	5

Like whole numbers, decimals can be written in expanded form.

Any number can be written in expanded form as the sum of decimals.

Example

Number	Expanded Form
3.5 =	(3 • 1) + (5 • 0.1)
18.34 =	(1 • 10) + (8 • 1) + (3 • 0.1) + (4 • 0.01)
48.675 =	(4 • 10) + (8 • 1) + (6 • 0.1) + (7 • 0.01) + (5 • 0.001)

Any decimal number can also be written in expanded form as the sum of fractions.

Example

Number	Expanded Form
3.5 =	$(3 \bullet 1) + (5 \bullet \frac{1}{10})$
18.34 =	$(1 \bullet 10) + (8 \bullet 1) + (3 \bullet \frac{1}{10}) + (4 \bullet \frac{1}{100})$
48.675 =	$(4 \bullet 10) + (8 \bullet 1) + (6 \bullet \frac{1}{10}) + (7 \bullet \frac{1}{100}) + (5 \bullet \frac{1}{1000})$

In summary:

Example

0.3

Place	Tenths
Number Means	3 tenths or $3 \bullet 0.1$
Value	0.3
In Words	Three-tenths
Fraction	$\frac{3}{10}$

0.06

Place	Hundredths
Number Means	6 hundredths or $6 \bullet 0.01$
Value	0.06
In Words	Six-hundredths
Fraction	$\frac{6}{100}$

Very Large and Very Small Numbers

In a number, you can have any number of places to the left and any number of places to the right of the decimal point.

Sometimes, numbers of large value are needed. For example, large numbers can be used to describe the population of a country or the distance to a planet or a star.

Sometimes, numbers of small value are needed. For example, small numbers can be used to describe the thickness of a strand of hair, the size of an atom, or the length of bacteria.

Example

431,678,854 is a number that represents a very large value.

Its value is:

4 hundred-millions + 3 ten-millions + 1 million + 6 hundred-thousands + 7 ten-thousands + 8 thousands + 8 hundreds + 5 tens + 4 ones

In words:

"Four hundred thirty-one million, six hundred seventy-eight thousand, eight hundred fifty-four"

In expanded form:

$(4 \bullet 100{,}000{,}000) + (3 \bullet 10{,}000{,}000) + (1 \bullet 1{,}000{,}000) + (6 \bullet 100{,}000) + (7 \bullet 10{,}000) + (8 \bullet 1{,}000) + (8 \bullet 100) + (5 \bullet 10) + (4 \bullet 1)$

Example

0.000435 is a number that represents a very small value.

Its value is:

0 ones + 0 tenths + 0 hundredths + 0 thousandths + 4 ten-thousandths + 3 hundred-thousandths + 5 millionths

In words:

"four hundred thirty-five millionths"

In expanded form as a sum of decimals:

$(0 \bullet 1) + (0 \bullet 0.1) + (0 \bullet 0.01) + (0 \bullet 0.001) + (4 \bullet 0.0001) + (3 \bullet 0.00001) + (5 \bullet 0.000001)$

The Base-10 System and Place Value

The Hindu-Arabic number system is a base-10 place-value system. This means it uses powers of ten, such as 1, 10, and 100, and that the *place* (or position) of each digit in a number determines that digit's value.

Example

In the number 705, the digit 7 has a value of 7 hundreds or 700, the digit 0 has a value of 0 tens or 0, and the digit 5 has a value of 5 ones or 5.

In 570, the digit 5 has a value of 5 hundreds or 500, and the digit 7 has a value of 7 tens or 70, and the digit 0 has a value of 0 ones or 0.

This table shows the first four whole-number places and the first two decimal places separated by a decimal point.

Thousands Place	Hundreds Place	Tens Place	Ones Place	.	Tenths Place	Hundredths Place
10 • 10 • 10 or 100 • 10	10 • 10	1 • 10	1	.	$\frac{1}{10}$	$\frac{1}{10 \cdot 10}$ or $\frac{1}{100}$

As you move to the left in the table, each place is ten times the value of the previous place. As you move, to the right in the table, each place is one tenth the value of the previous place. The places in the place-value table are the powers of ten: 1, 10, 100 $\frac{1}{10}$, $\frac{1}{100}$ and so on.

In the number system, there can be any number of place-value columns to the left and to the right of the decimal point.

THE NUMBER LINE

CHAPTER 5

Number Lines

In many ways a number line is like a ruler. Any number can be located on the number line. A positive number is equal to its distance from 0 on the line.

2 is twice as far from 0 as 1.
3 is farther from 0 than 2.

On a number line, numbers always increase as you move to the right. They decrease as you move to the left.

Here is part of a number line, from 0 to 14. The arrows at each end of the line mean that the number line goes on forever in both directions.

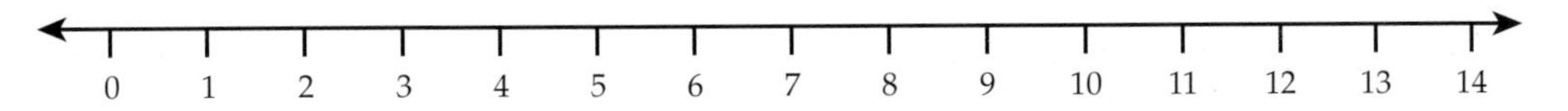

You can show, and use, any part of the number line. You do not have to show the part with 0 on it. Here is another part of a number line. You know this number line is to the right of the number line above because numbers increase as you go to the right.

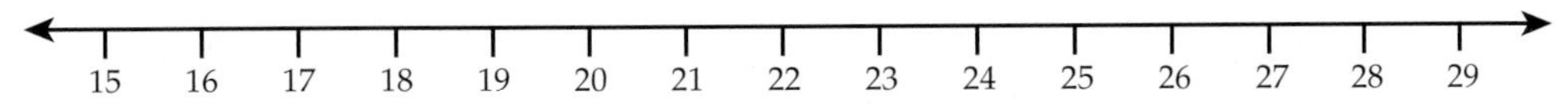

Both number lines show whole numbers in intervals of one. Each number is one greater than the number to its left. Each number is one less than the number directly to its right.

Consecutive numbers on a number line are always an equal distance apart. The distance between marks is called an *interval*. On some number lines, consecutive numbers may be shown far apart but still have an equal distance between numbers.

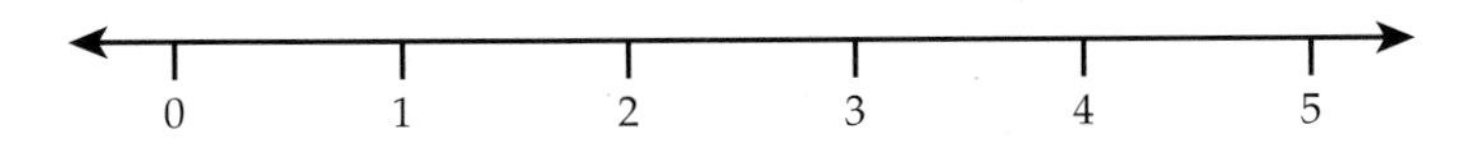

On some number lines, the numbers may be shown close together, but must still be at an equal distance.

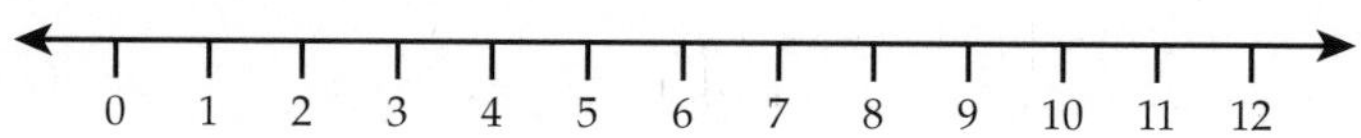

The example below is not a correct number line. The numbers are in order, but the numbers are not placed an equal distance apart.

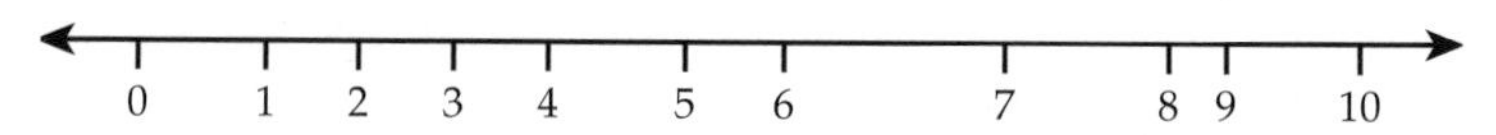

Locating Whole Numbers on Number Lines

Every number has a location, or place, where it belongs on the number line. For a positive number, the location is its distance to the right from 0. To locate numbers on a number line, you count the intervals from 0, just like on a ruler.

Example

The point on the number line below shows the location of the number 3. The number 3 is three units, or three intervals, to the right of 0. It is between the numbers 2 and 4.

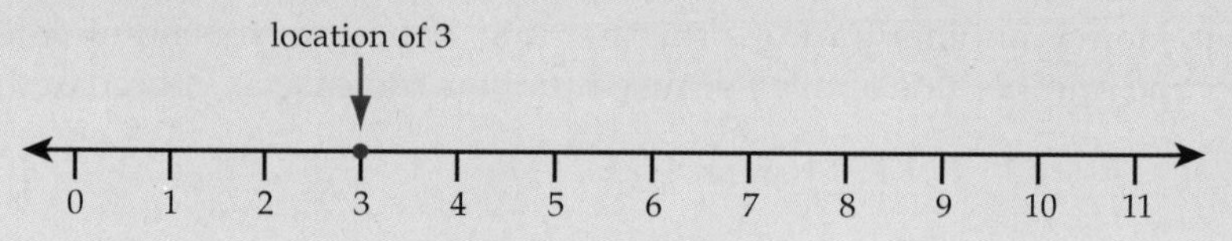

Example

This number line shows whole numbers from 20 to 34.
The point shows the location of the number 27.

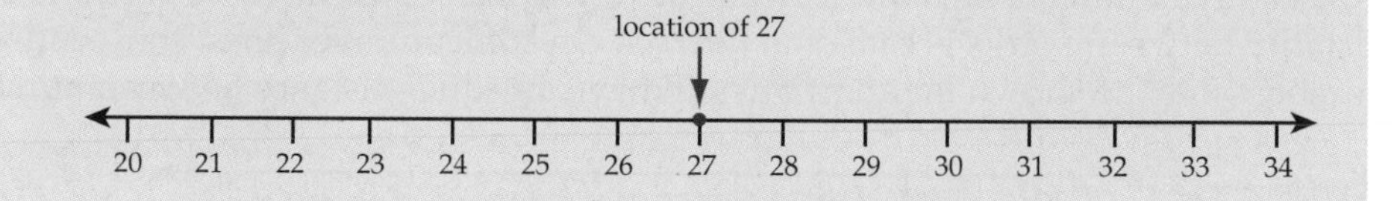

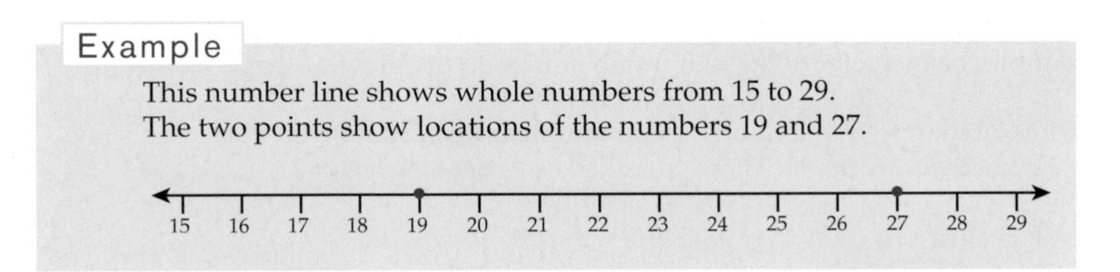

Locating Fractions on Number Lines

You can show a close-up of the number line, so that very small intervals between the numbers can be seen.

Here are four number lines that show close-ups of the number line between 0 and 1. Each is divided into a different number of intervals.

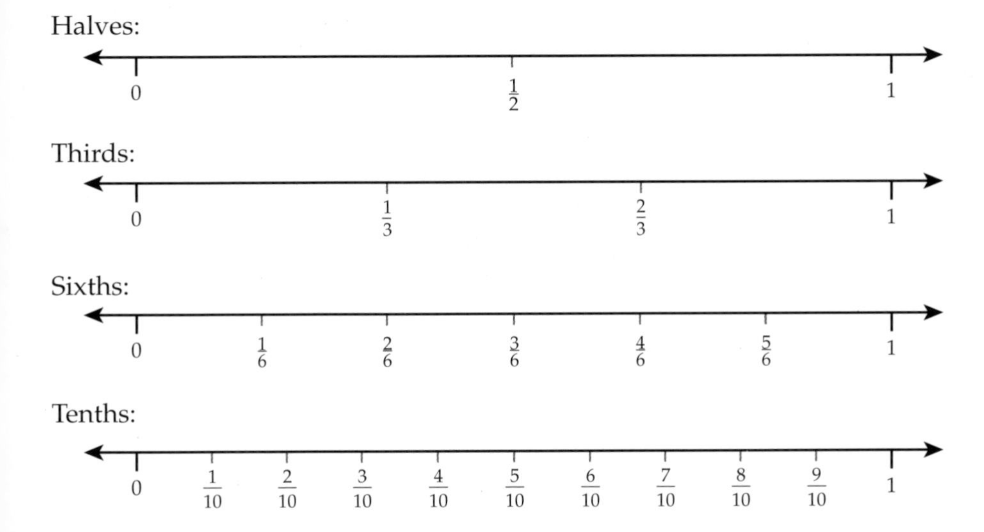

Number lines are useful for comparing fractions. When number lines are shown in a column, with equivalent numbers lined up (as the four number lines shown above), you can place a ruler perpendicular to the number lines to compare fractions.

You can use number lines to compare numbers. In working through the three examples below, refer to the four number lines on the previous page.

1. Comparing fractions on different number lines allows you to see that some fractions are equal. These are called *equivalent fractions*.

Example

$\frac{1}{2} = \frac{3}{6} = \frac{5}{10}$ $\frac{2}{3} = \frac{4}{6}$

2. Comparing fractions on different number lines also allows you to see midpoints.

Example

$\frac{3}{6}$ is the midpoint between $\frac{1}{3}$ and $\frac{2}{3}$.

$\frac{3}{6}$ is also the midpoint between $\frac{4}{10}$ and $\frac{6}{10}$.

3. Comparing fractions on different number lines allows you to see size relationships between numbers.

Example

$\frac{1}{3}$ is greater than $\frac{2}{10}$.

$\frac{1}{6}$ is between $\frac{1}{10}$ and $\frac{2}{10}$, but it is closer to $\frac{2}{10}$.

Number Lines and Decimals

Number lines can show decimals. You can move closer (zoom in) to show smaller intervals or move away (zoom out) to show larger intervals on the number line.

Here is a number line, from 0 to 1, that is divided into tenths. This is a good way to locate and compare decimal numbers from 0 to 1, because each division on the number line represents $\frac{1}{10}$ or 0.1. Tenths are the first decimal place.

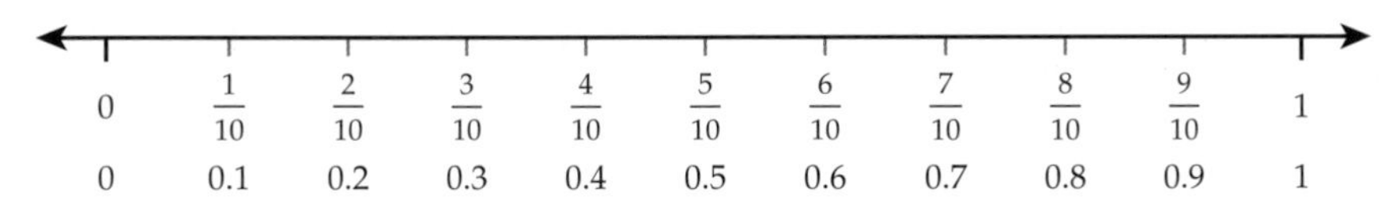

Looking at the number line above, you can see many different relationships.

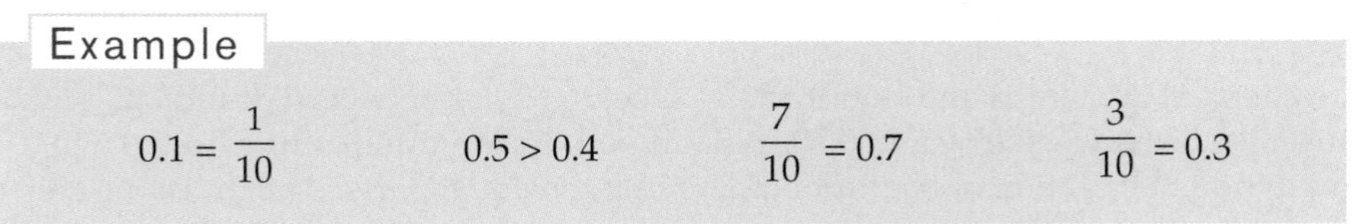

Example

$0.1 = \frac{1}{10}$ $\quad 0.5 > 0.4$ $\quad \frac{7}{10} = 0.7$ $\quad \frac{3}{10} = 0.3$

If the space from 0 to 0.1 is divided into ten equal parts, then each division on the number line represents 0.01 or $\frac{1}{100}$. Hundredths are in the second decimal place.

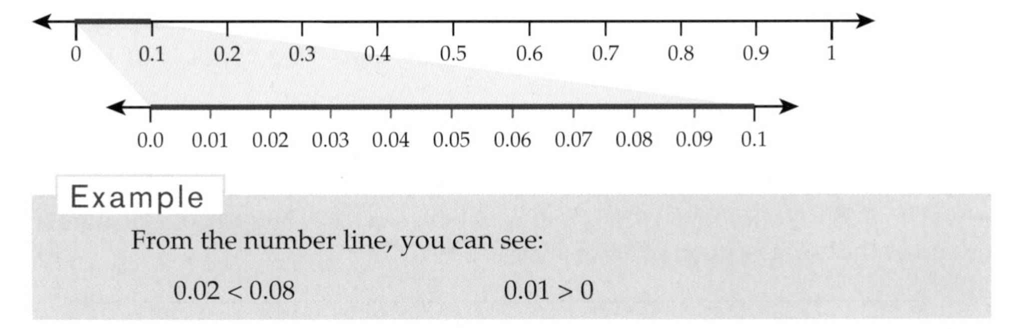

Example

From the number line, you can see:

$0.02 < 0.08$ $\quad 0.01 > 0$

You can continue to expand an interval and divide it into ten new intervals forever! Every time you do this, you are representing the next decimal place to the right.

Locating Decimals on Number Lines

Suppose you want to locate 0.625 on a number line.

First, you need to decide which section of the number line to represent.
How many divisions would be best?

Look at the ones and tenths places in 0.625. The number 0.6 lies between 0.0 and 1.0, so 0.625 must also lie somewhere between 0.0 and 1.0. What if you draw a number line from 0 to 1 and divide the number line into ten equal parts? The divisions are labeled 0, 0.1, and so on, all the way to 1. It is possible to locate 0.625 on this number line approximately; the decimal number 0.625 lies somewhere between 0.6 and 0.7.

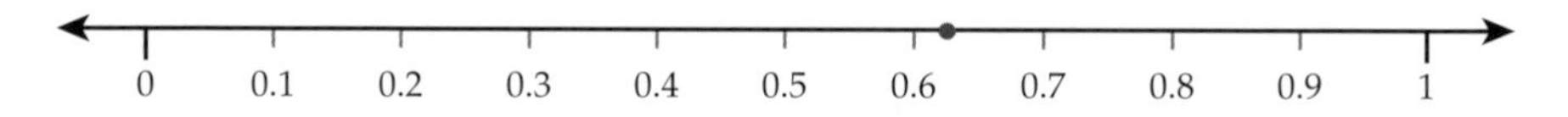

Now, look at the tenths and hundredths places. 0.62 lies between 0.6 and 0.7, so 0.625 must also lie between 0.60 and 0.70. What if you draw a number line from 0.6 to 0.7 and divide that section of the number line into ten equal parts? The divisions are labeled 0.6, 0.61, and so on, all the way to 0.7.

You can locate 0.625 on the number line. 0.625 lies between 0.62 and 0.63. However, you still do not have an exact location.

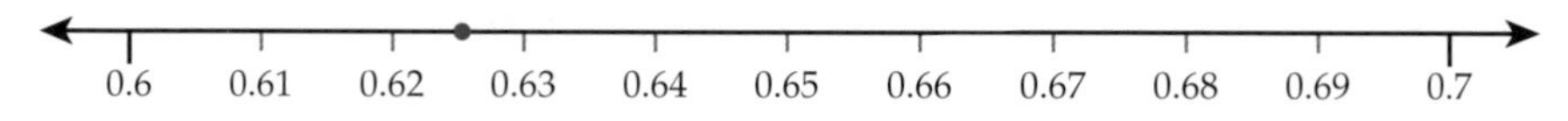

Now, look at the thousandths place. 0.625 lies between 0.620 and 0.630. What if you draw a number line from 0.620 to 0.630 and divide that section of the number line into ten equal parts? Now, the divisions are labeled 0.620, 0.621, 0.622, and so on, all the way to 0.630. Notice that 0.625 is half way between 0.620 and 0.630. This number line shows the exact location of 0.625.

Finding Numbers between Two Numbers

Integers are the numbers …, –3, –2, –1, 0, 1, 2, 3, 4, 5,…

Consecutive integers are two integers that are next to each other when the integers are listed in order from least to greatest.

Example

3 and 4 are consecutive integers.

3 and 5 are not consecutive integers, because 4 is between 3 and 5.

It is not possible to find an integer between consecutive integers.

Example

There is no integer between 3 and 4.

It is possible to find a rational number between consecutive integers.

Example

The rational numbers 3.1, 3.5, 3.7, and many more, are between 3 and 4.

The symbols, <, "is less than" and >, "is greater than," are used to show numbers that have a value between two numbers.

Example

In Symbols	In Words
$3 < 3.1 < 4$	"3 is less than 3.1 and 3.1 is less than 4."
$4 > 3.1 > 3$	"4 is greater than 3.1 and 3.1 is greater than 3."

It is always possible to find a rational number that has a value between two other rational numbers.

Example

Find a rational number between 3.1 and 3.2:

3.14, 3.18, 3.19 are between 3.1 and 3.2

You can show this as:

$$3.1 < 3.14 < 3.18 < 3.19 < 3.2$$

You can "zoom in" on the number line to show these numbers.

Example

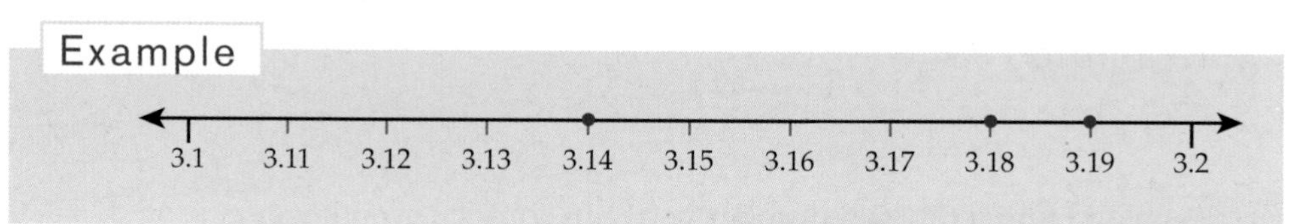

ADDITION AND SUBTRACTION

CHAPTER 6

The Concept of Addition

Addition and subtraction are mathematical operations.

Addition can be used when combining two or more groups of objects.
The symbol "+" represents the operation "*add*." This symbol is often read as "*plus*."

Look at this representation of addition.

+ =

9 flags + 7 flags = 16 flags

Numbers that are added are called *addends*. The result of an addition is called the *sum* or *total*.

Example

$9 + 7 = 16$ The addends are 9 and 7.

The sum, or total, is 16.

Chapter 6

The Concept of Subtraction

Subtraction can be used when removing or taking away part of a group of objects. The symbol "–" represents the operation "*subtract*." This symbol is often read as "*minus*" or "*take away*."

Look at this representation of subtraction.

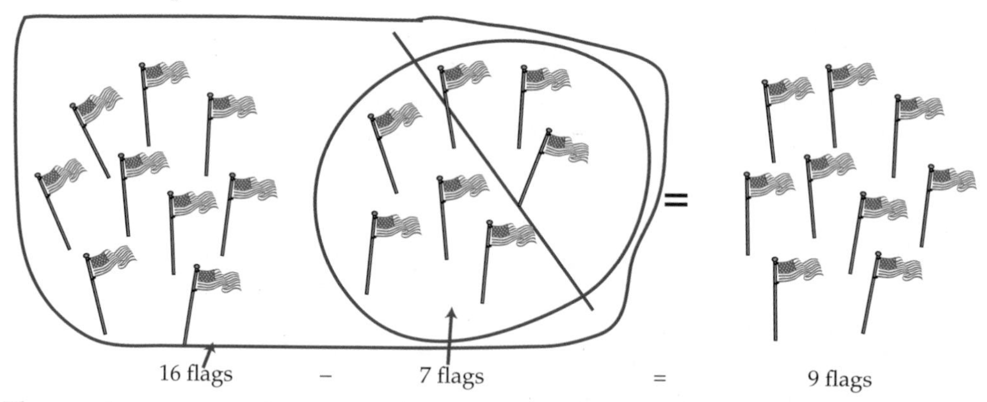

The number you get when you subtract one number from another number is the *difference* between the two numbers. In $16 - 7 = 9$, the difference between 16 and 7 is 9.

Here is another way to represent this subtraction. Imagine that you have a 16-centimeter piece of ribbon. Think of each centimeter of the ribbon as one unit on the number line. If you have a 16-centimeter piece of ribbon and you cut 7 centimeters off, you will have 9 centimeters left.

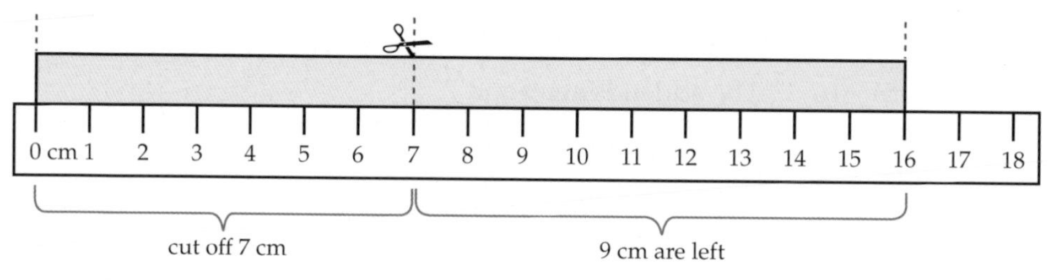

The equation that represents this subtraction is $16 - 7 = 9$.

Imagine you have another 16-centimeter piece of ribbon. If you cut 9 centimeters off, you will have 7 centimeters left.

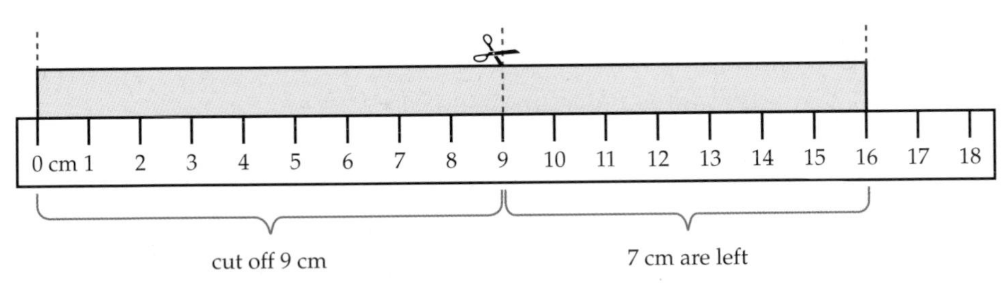

The equation that represents this subtraction is 16 – 9 = 7. The difference between 16 and 9 is 7, and the difference between 16 and 7 is 9.

Addition and Subtraction on the Number Line

Addition on the Number Line

You can use number lines to "see" addition.

Example

Look at this number line representing 7 + 3 = 10.

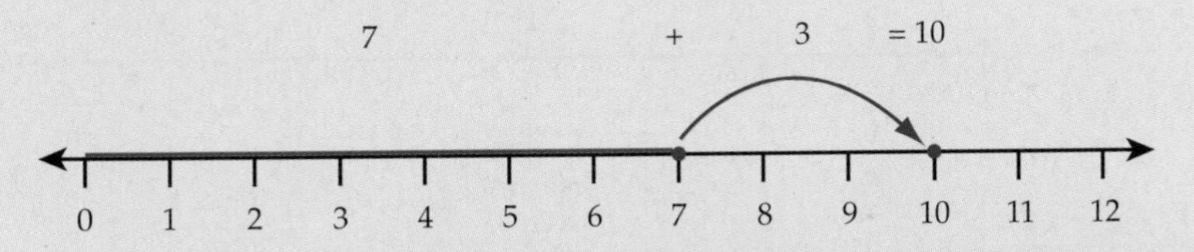

A point shows the location of 7 on the number line. You can think of the addition "7 + 3" as meaning "add a distance of 3 to 7."

The point on the 10 shows the place on the number line that is 3 units more than 7.

The sum, 10, is 3 more than the starting number, 7.

Chapter 6

Subtraction on the Number Line

You can use number lines to "see" subtraction.

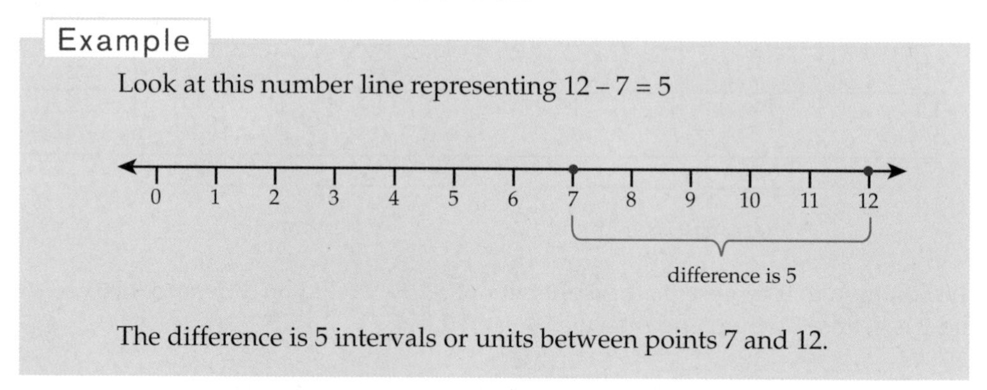

Distance on the Number Line

Number lines can also help you see or think about distance. The distance between two numbers on a number line tells you their difference.

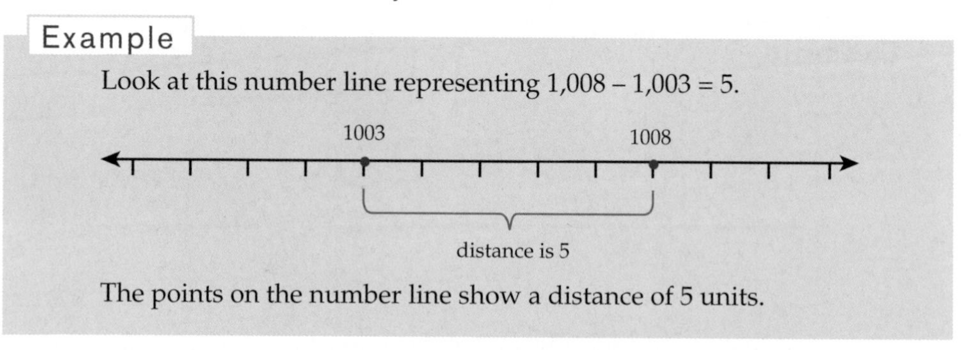

A distance of 5 units could show a difference between 15 and 20, between 1,003 and 1,008, or between any two numbers that are 5 units apart.

Addition and Subtraction Are Inverse Operations

Addition and subtraction are *inverse operations*; they undo each other.

Example

If you have 7 pencils and I take away 5 pencils, then you have 2 pencils: $7 - 5 = 2$.

If I give you back 5 pencils, then once again you have 7 pencils: $2 + 5 = 7$.

Subtracting 5 pencils and adding 5 pencils are inverse operations. They undo each other. If you start with 7, then subtract 5, then add 5, you return to 7.

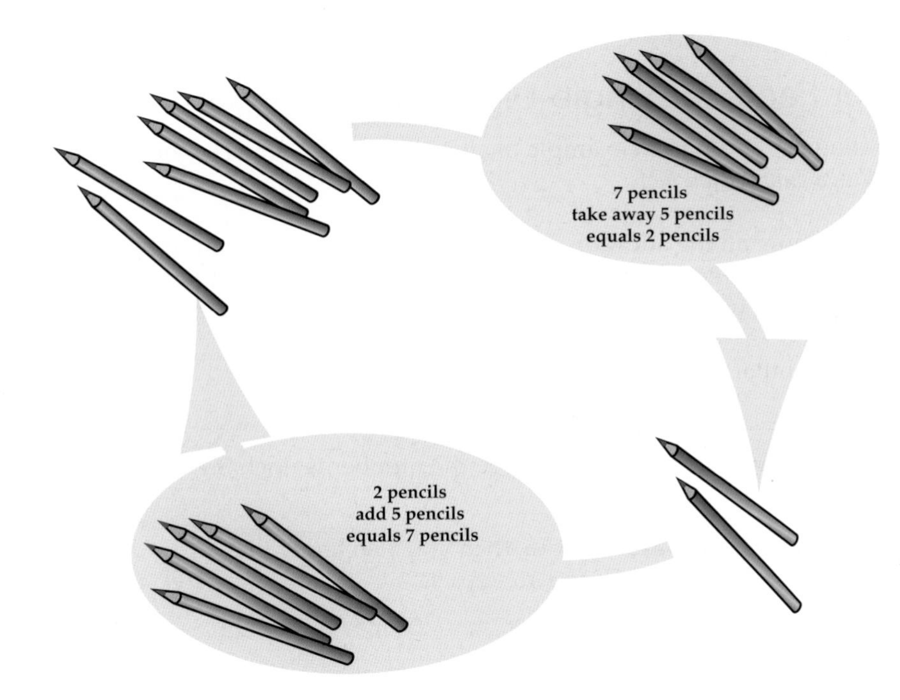

You can think of moving to the right on the number line as adding (counting on) and moving to the left as subtracting (counting back).

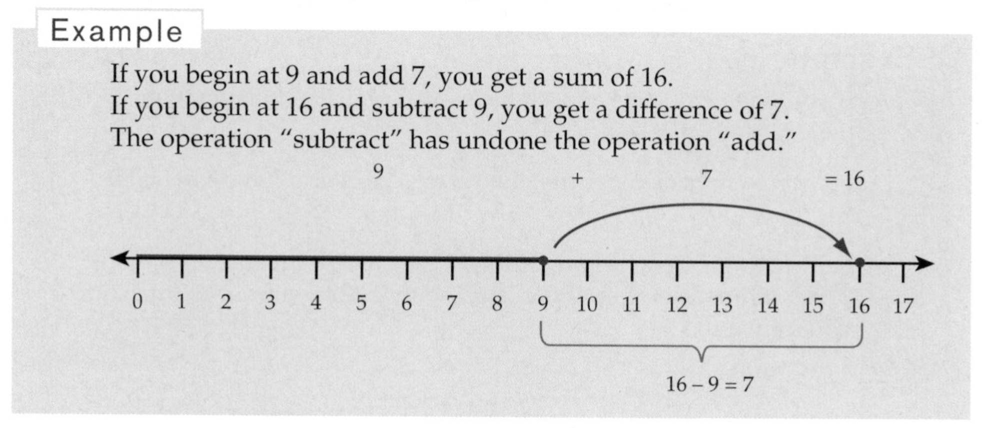

Number Facts for Single-Digit Numbers

The equation $9 + 7 = 16$ is an example of an *addition fact*. Similarly, the equation $16 - 9 = 7$ is a *subtraction fact*.

You need to know all the addition facts for all pairs of single-digit numbers by heart. This knowledge will enable you to perform more difficult calculations quickly and efficiently without using a calculator.

Some of the number facts are easy to remember.

Example

If you add zero to any number, the number is unchanged. This means $1 + 0 = 1$, and so on.

If you know that $9 + 7 = 16$, then you also know that $7 + 9 = 16$, $16 - 9 = 7$, and $16 - 7 = 9$.

At the bottom of this page, you will find a table of some of the addition facts. Note that many of the entries are related because addition is commutative.

Example

Since $8 + 9 = 9 + 8$, the number fact $9 + 8 = 17$ is related to the number fact $8 + 9 = 17$.

Also, since addition and subtraction are inverse operations, knowing one number fact means that you know three other number facts. If you know the addition facts, then you have enough information to figure out the subtraction facts, and vice versa.

Example

From $8 + 9 = 17$, you know the subtraction facts

$$17 - 8 = 9 \text{ and } 17 - 9 = 8.$$

Subtraction is not commutative.

Example

$2 + 6 = 6 + 2$, but $6 - 2 \neq 2 - 6$.

Addition Facts

1 + 1 = 2	2 + 1 = 3	3 + 1 = 4	4 + 1 = 5	5 + 1 = 6	6 + 1 = 7	7 + 1 = 8	8 + 1 = 9	9 + 1 = 10
1 + 2 = 3	2 + 2 = 4	3 + 2 = 5	4 + 2 = 6	5 + 2 = 7	6 + 2 = 8	7 + 2 = 9	8 + 2 = 10	9 + 2 = 11
1 + 3 = 4	2 + 3 = 5	3 + 3 = 6	4 + 3 = 7	5 + 3 = 8	6 + 3 = 9	7 + 3 = 10	8 + 3 = 11	9 + 3 = 12
1 + 4 = 5	2 + 4 = 6	3 + 4 = 7	4 + 4 = 8	5 + 4 = 9	6 + 4 = 10	7 + 4 = 11	8 + 4 = 12	9 + 4 = 13
1 + 5 = 6	2 + 5 = 7	3 + 5 = 8	4 + 5 = 9	5 + 5 = 10	6 + 5 = 11	7 + 5 = 12	8 + 5 = 13	9 + 5 = 14
1 + 6 = 7	2 + 6 = 8	3 + 6 = 9	4 + 6 = 10	5 + 6 = 11	6 + 6 = 12	7 + 6 = 13	8 + 6 = 14	9 + 6 = 15
1 + 7 = 8	2 + 7 = 9	3 + 7 = 10	4 + 7 = 11	5 + 7 = 12	6 + 7 = 13	7 + 7 = 14	8 + 7 = 15	9 + 7 = 16
1 + 8 = 9	2 + 8 = 10	3 + 8 = 11	4 + 8 = 12	5 + 8 = 13	6 + 8 = 14	7 + 8 = 15	8 + 8 = 16	9 + 8 = 17
1 + 9 = 10	2 + 9 = 11	3 + 9 = 12	4 + 9 = 13	5 + 9 = 14	6 + 9 = 15	7 + 9 = 16	8 + 9 = 17	9 + 9 = 18

Chapter 6

Mental Strategies for Adding and Subtracting

For many calculations, you can use your knowledge of place value and number facts to figure out results in your head.

- These addition facts are particularly useful, because each sum is ten.

Example

$1 + 9 = 10$	$3 + 7 = 10$	$5 + 5 = 10$	$7 + 3 = 10$	$9 + 1 = 10$
$2 + 8 = 10$	$4 + 6 = 10$	$6 + 4 = 10$	$8 + 2 = 10$	$10 + 0 = 10$

Be on the lookout for additions such as $9 + 8 + 1$. If you first add 9 and 1 to get 10, then you can easily see that the answer is $10 + 8 = 18$.

- It is easy to add or subtract numbers to multiples of ten and powers of ten.

Example

$5 + 10 = 15$	$100 + 19 = 119$	$3{,}000 + 69 = 3{,}069$

Example

$546 - 500 = 46$	$2{,}567 - 100 = 2{,}467$	$3{,}345 - 300 = 3{,}045$

- Use mental regrouping so you can add multiples of ten and powers of ten.

Example

$36 + 67 = (36 + 4) + (67 - 4) = 40 + 63 = 103$
$99 - 12 = (99 + 1) - 12 - 1 = 100 - 13 = 87$

- Always look to see where you can use number facts to solve additions in your head.

Example

Think about the addition fact $9 + 7 = 16$. Whenever 7 is added to a number ending in 9, the result is always a number ending in 6. You can use this fact to carry out additions like $209 + 7$.

$$209 + 7 = 200 + 9 + 7 = 200 + 16 = 216$$

- Always look to see where you can use number facts to solve subtractions in your head.

Example

When 7 is subtracted from a number ending in 6, the result is always a number ending in 9.

The number fact that this is based on is $16 - 7 = 9$.

$$26 - 7 = 19$$

The result, 19, is 10 more than 9, because 26 is ten more than 16.

Addition with Regrouping and Base-10 Blocks

Our number system is based on the powers of ten. You can use this fact when adding.

Example

Think about the addition 5 + 11.

This is 5 + 10 + 1

Group the ones: (5 + 1 = 6)

Then you have 1 ten and 6 ones , which makes 16.

Example

When you add 7 and 9, the result is more than 10. This means that a group of 10 must be made from the combined group, leaving 6 ones.

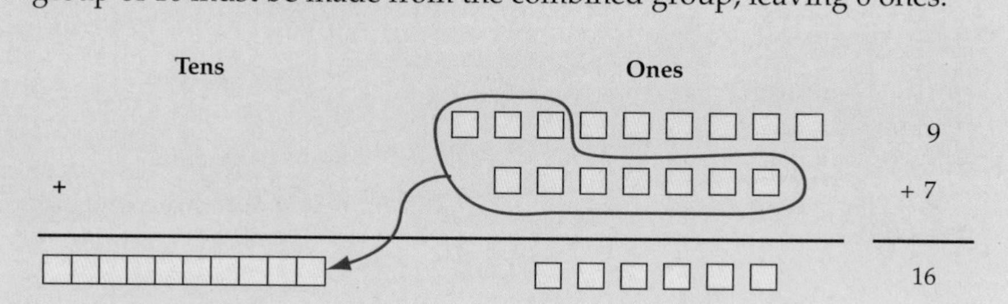

The digit 1 is written in the tens place and the digit 6 is written in the ones place. This process of combining numbers into groups of ten is called *regrouping*.

The place value notation "16" represents the result of the addition "9 + 7" after regrouping has occurred.

Example

What regrouping needs to happen when 59 is added to 43?

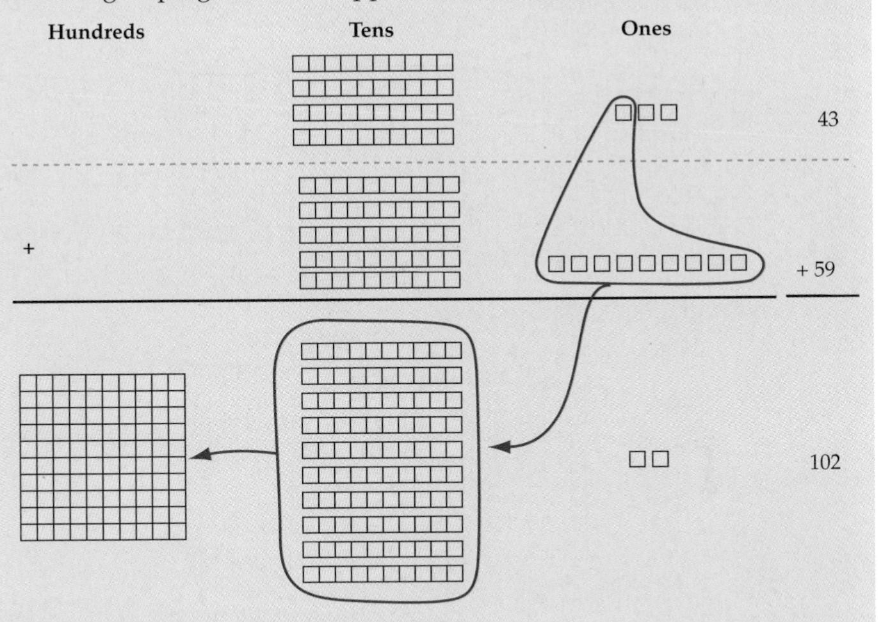

When adding the ones, the twelve ones are regrouped into 1 ten and 2 ones, then the 10 tens are regrouped to make 1 hundred. The total is 102.

Subtraction with Regrouping and Base-10 Blocks

Our number system is based on the powers of ten. You can use this fact when subtracting.

Example

45 – 32

This difference is easy to calculate, because 5 – 2 = 3 ones and 4 – 3 = 1 ten. Thus 45 – 32 = 13.

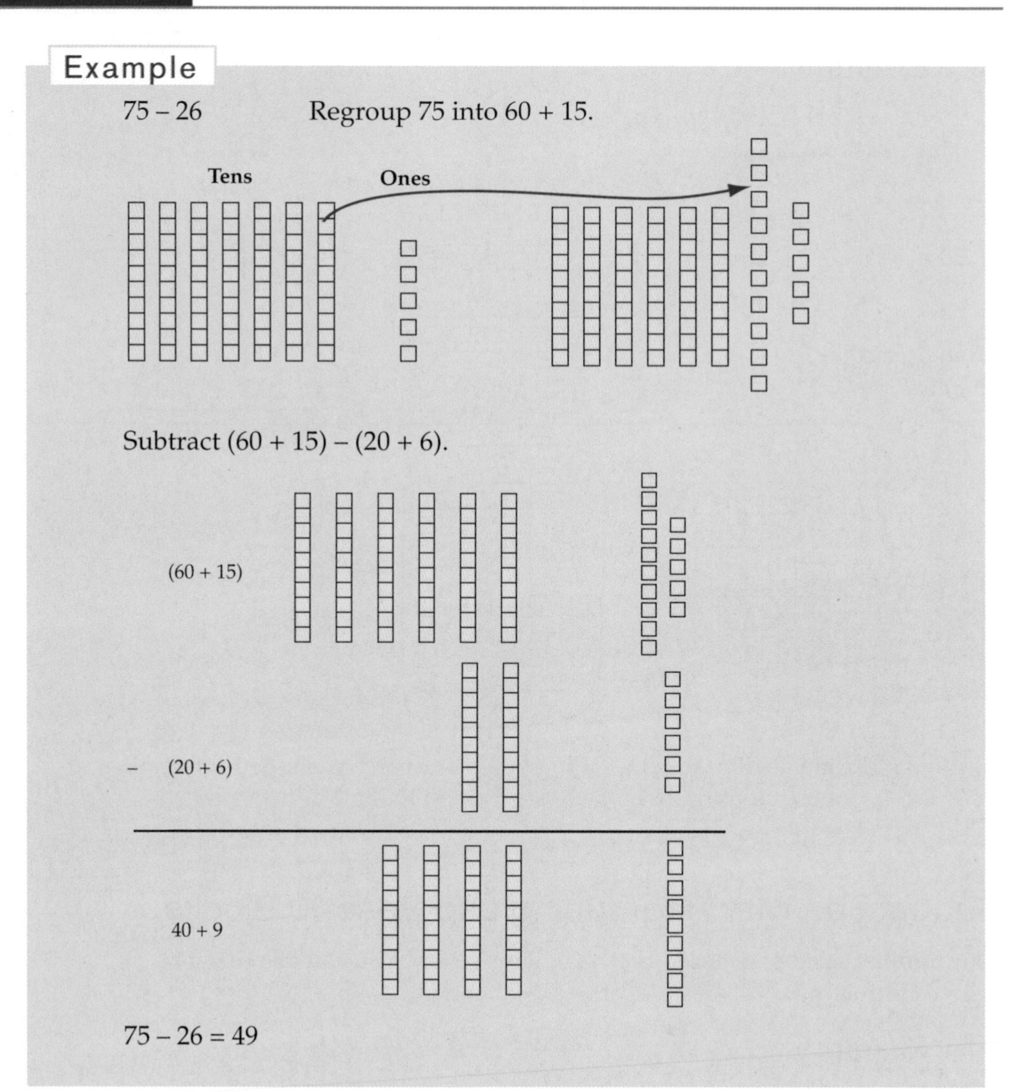
Example
75 – 26
Regroup 75 into 60 + 15.
Tens
Ones
Subtract (60 + 15) – (20 + 6).
(60 + 15)
– (20 + 6)
40 + 9
75 – 26 = 49

Written Methods for Addition and Subtraction

It is important that you understand and learn how to add and subtract without a calculator.

When you are comfortable with all of the methods of adding and subtracting, you will not need a calculator as much. You can check your answers using a calculator.

Before you perform any addition or subtraction calculation, you should:

- carefully consider the place value of the digits in the numbers, and
- make an estimate of the result.

When you use written methods to add and subtract, it is important that the digits are written in the correct places, with the ones under the ones, the tens under the tens, and so on.

The Standard Method for Addition

The standard method for addition is based on regrouping.

Remember, always estimate the total first.

Example

65 + 138

An estimate is 60 + 140 = 200.

		Hundreds / Tens / Ones
Add the ones	5 + 8 = 13 ones	
Regroup	13 ones = (1 • 10) + (3 • 1) The 1 digit belongs in the tens place. "Carry" the 1 to the tens place. Write the 3 in the ones place.	$\begin{array}{r} {}^{1} \\ 65 \\ +138 \\ \hline 3 \end{array}$

		Hundreds / Tens / Ones
Add the tens	6 + 3 + 1 = 10 tens	
Regroup	10 tens = (1 • 100) + (0 • 10) "Carry" the 1 to the hundreds place. Write the digit 0 in the tens place.	$\begin{array}{r} {}^{1}\,{}^{1} \\ 65 \\ +138 \\ \hline 03 \end{array}$

		Hundreds / Tens / Ones
Add the hundreds	1 + 1 = 2 hundreds The digit 2 belongs in the hundreds place. Write the digit 2 in the hundreds place.	$\begin{array}{r} {}^{1}\,{}^{1} \\ 65 \\ +138 \\ \hline 203 \end{array}$

Write as an equation, 65 + 138 = 203.

The Standard Method for Subtraction

You can use the standard method and regrouping to solve subtraction problems.

Example

403 – 87 Estimate: 400 – 90 = 310

		Hundreds Tens Ones
Regroup the ones column	You cannot subtract 7 ones from 3 ones. You need to regroup 1 ten from the tens place into 10 ones to make 13 ones, but there are 0 tens! Regroup 1 hundred into 10 tens, and then 1 ten into 10 ones. Subtract 7 from 13. Put a 6 in the ones place.	$\begin{array}{r} \overset{3}{\cancel{4}}\ \overset{9}{\cancel{0}}\ {}^{1}3 \\ -\ \ 8\ \ 7 \\ \hline 6 \end{array}$
Subtract the tens	Subtract 8 from 9 to get 1. Put a 1 in the tens place.	$\begin{array}{r} \overset{3}{\cancel{4}}\ \overset{9}{\cancel{0}}\ {}^{1}3 \\ -\ \ 8\ \ 7 \\ \hline 1\ \ 6 \end{array}$
Subtract the hundreds	Subtract 0 from 3 to get 3. Put a 3 in the hundreds place.	$\begin{array}{r} \overset{3}{\cancel{4}}\ \overset{9}{\cancel{0}}\ {}^{1}3 \\ -\ \ 8\ \ 7 \\ \hline 3\ \ 1\ \ 6 \end{array}$

Write as an equation, 403 – 87 = 316.

Chapter 6

Addition and Subtraction of Decimals

All the concepts of addition and subtraction apply to decimals, because decimal numbers are a continuation of the base-10 number system.

The rules of arithmetic work for addition and subtraction of decimals, just as they do for whole numbers.

- The addition facts can be used.

Example

0.6 + 0.8 = 1.4 and 0.3 + 0.7 = 1.0
You know the addition facts 6 + 8 = 14 and 3 + 7 = 10.
The number line will also show the additions:

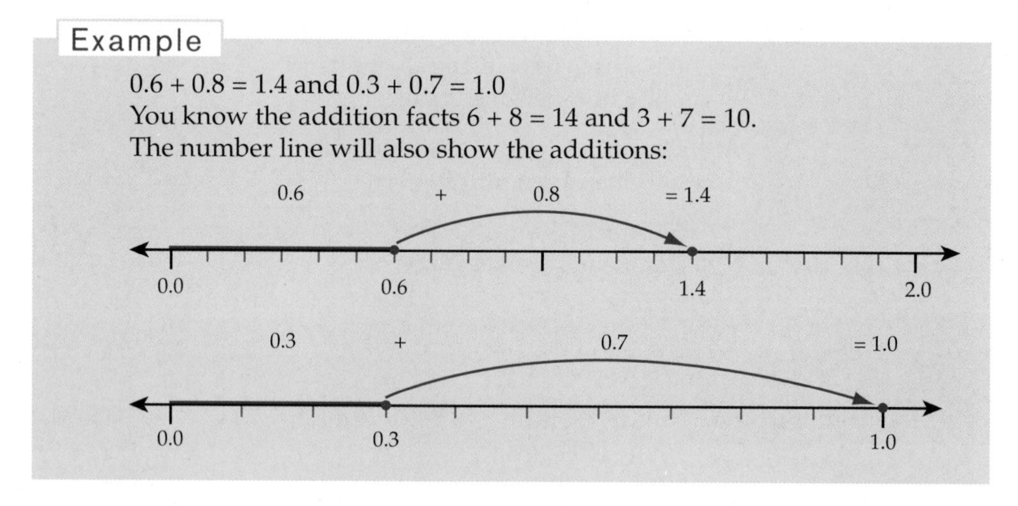

- Adding decimals in any order gives the same total. As with all numbers, adding decimals is commutative.

Example

0.6 + 0.1 = 0.1 + 0.6 = 0.7

- Addition and subtraction of decimals are inverse operations.

Example

Since 0.4 + 0.7 = 1.1, then 1.1 – 0.7 = 0.4 and 1.1 – 0.4 = 0.7.

- Adding zero to a decimal number does not change its value.

Example

Since 0 + 0.4 = 0.4, then 0.0 + 0.4 = 0.4.

- You add and subtract decimals the same way as you do whole numbers. Just as with whole numbers, you need to make sure the numbers, in this case the decimal places, are lined up correctly.

Example

1.89 + 0.37 can be calculated by regrouping using the standard method for addition.

$$\begin{array}{rlll} & \text{Ones} & \text{Tenths} & \text{Hundredths} \\ & {}^{1} & {}^{1} & \\ & 1. & 8 & 9 \\ + & 0. & 3 & 7 \\ \hline & 2. & 2 & 6 \end{array}$$

Example

4.56 – 1.99 can be calculated by regrouping using the standard method for subtraction.

$$\begin{array}{rlll} & \text{Ones} & \text{Tenths} & \text{Hundredths} \\ & {}^{3} & {}^{14} & \\ & \not{4}. & \not{5} & {}^{1}6 \\ - & 1. & 9 & 9 \\ \hline & 2. & 5 & 7 \end{array}$$

Chapter 6

Addition and Subtraction of Fractions

Simple fractions with the same denominator can be added and subtracted, just like whole numbers and decimals.

Example

$\frac{4}{7} + \frac{2}{7} = \frac{6}{7}$ and $\frac{5}{7} + \frac{6}{7} = \frac{11}{7}$

You can see these equations on the number line.

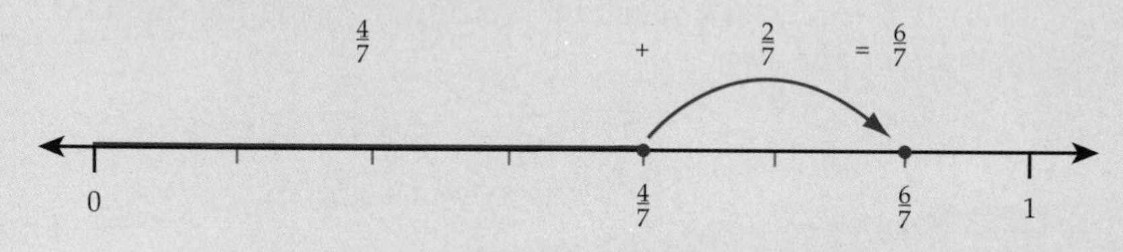

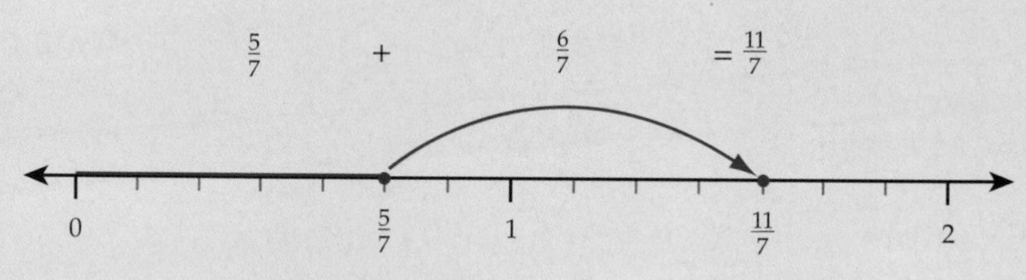

Notice that the fraction $\frac{11}{7}$ can be regrouped to make $\frac{7}{7} + \frac{4}{7} = 1\frac{4}{7}$.

The rules of arithmetic hold true for addition and subtraction of fractions, just as they do for whole numbers and decimals.

- Adding fractions in any order gives the same total.
  Adding fractions is commutative.

Example

$$\frac{4}{7} + \frac{2}{7} = \frac{2}{7} + \frac{4}{7}$$

- Addition and subtraction of fractions are inverse operations.

Example

$\frac{4}{7} + \frac{2}{7} = \frac{6}{7}$ and $\frac{6}{7} - \frac{2}{7} = \frac{4}{7}$ and $\frac{6}{7} - \frac{4}{7} = \frac{2}{7}$

- Adding zero to a fraction does not change its value.

Example

Since $\frac{4}{7} + 0 = \frac{4}{7}$, then $\frac{4}{7} + \frac{0}{7} = \frac{4}{7}$.

The Commutative Property of Addition

It is always true that for any two numbers a and b, $a + b = b + a$.
This is called the *commutative property of addition*.

Example

The sum of 9 and 7 is 16, no matter the order in which they are added.

You can show this using equivalent equations:

$$9 + 7 = 16 \qquad 7 + 9 = 16$$
$$9 + 7 = 7 + 9$$

Example

Let $a = 12$ and $b = 15$.

$$a + b = b + a$$
$$12 + 15 = 15 + 12$$
$$12 + 15 = 27 \qquad 15 + 12 = 27$$

There is a very important difference between addition and subtraction: the commutative property does not work for subtraction.

Example

$9 - 7$ equals 2, but $7 - 9$ does not equal 2.

$9 - 7 \neq 7 - 9$

Subtraction is different from addition in an important way: the order of *adding* two numbers does not change the result, but the order of *subtracting* one number from another does change the result.

MULTIPLICATION AND DIVISION

CHAPTER 7

The Concept of Multiplication

Multiplication gives the *product* of two or more numbers and can be shown:

Using the symbol ×, as in $5 \times 4 = 20$

Using the symbol •, as in $5 \bullet 4 = 20$

With no symbol at all, when used with letters

Example

$5y = 20$ (meaning $5 \bullet y = 20$) $xy = 20$ (meaning $x \bullet y = 20$)

In any multiplication, the numbers you are multiplying are called the *factors*, and the result is called the *product*.

Example

$5 \bullet 4 = 20$ The factors are 4 and 5. The product is 20.

Think of multiplication as a quick way to add a given number of equal groups.

Example

Suppose you need to add three groups of four: $4 + 4 + 4$. The number line shows this repeated addition has a sum of 12.

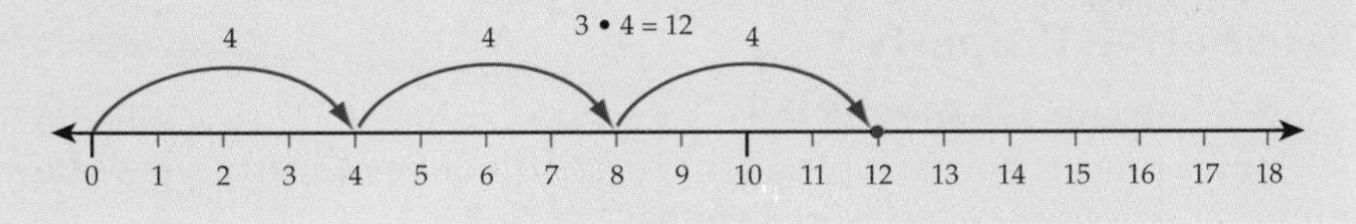

The multiplication $3 \bullet 4$ is a short representation of the repeated addition:

$$3 \bullet 4 = 4 + 4 + 4 = 12$$

Example

Suppose you need to add four groups of three: 3 + 3 + 3 + 3. The number line shows this repeated addition has a sum of 12.

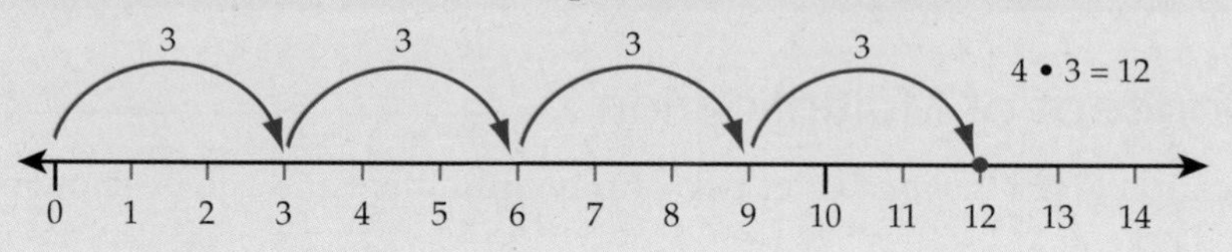

These two examples, 3 • 4 = 12 and 4 • 3 = 12, illustrate the *commutative property of multiplication*. For any numbers a and b, $a \bullet b = b \bullet a$.

Multiplication Represented on the Number Line

Multiplication of decimals can also be represented as repeated addition on the number line.

Example

The multiplication 1.4 • 4 gives the product 5.6.

This calculation can be represented on the number line, where 1.4 is the size of each step and 4 is the number of steps, starting at zero.

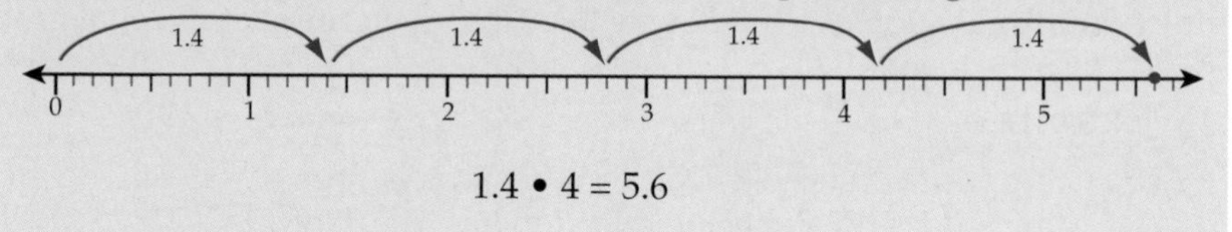

1.4 • 4 = 5.6

Distributive Property

For any numbers a, b, and c, $a(b + c) = a \bullet b + a \bullet c$. This property is called the distributive property of multiplication. The distributive property shows how addition and multiplication work together.

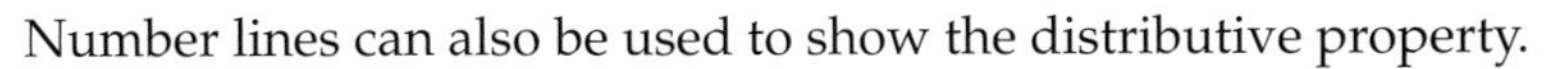

Number lines can also be used to show the distributive property.

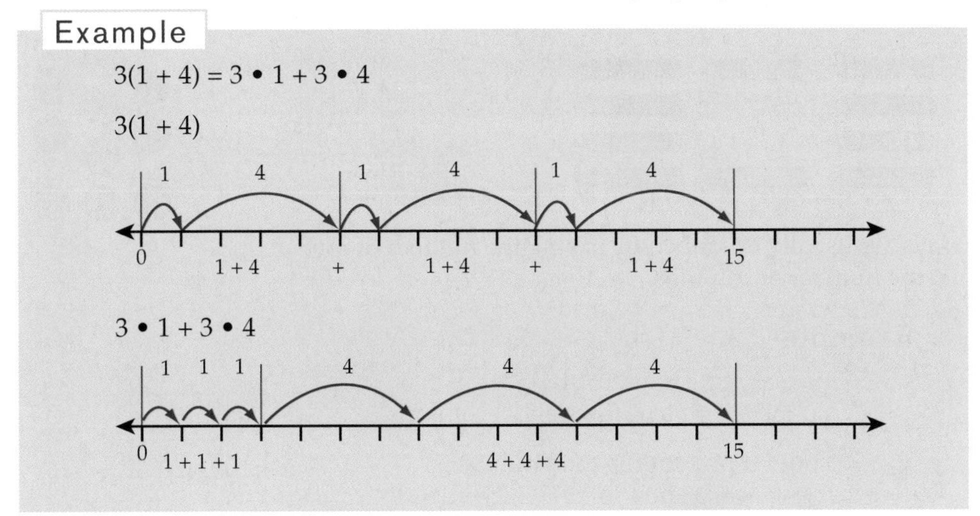

Multiplication Represented in Arrays

A *rectangular array* (array for short) is an arrangement of objects in rows and columns. Arrays can be made out of any number of objects that can be put into equal numbers of rows and columns.

> *Rows* are horizontal. They go from side to side. *Columns* are vertical. They go from top to bottom. In an array, all rows contain an equal number of items and all columns contain an equal number of items.

These are three different arrays:

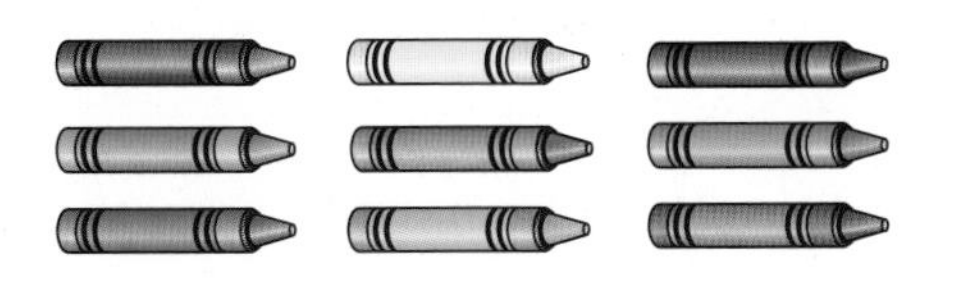

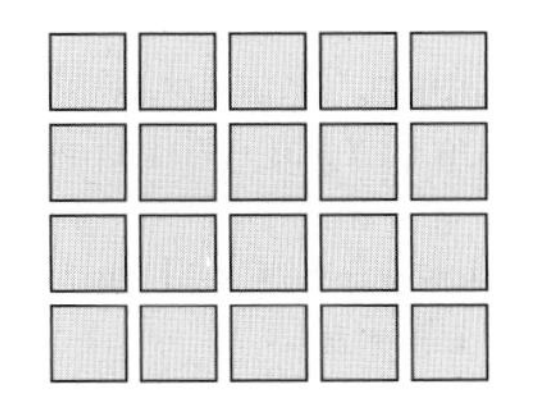

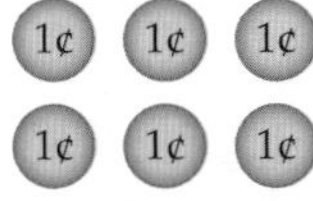

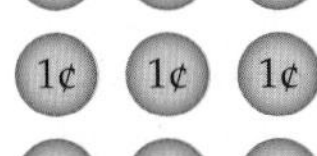

Because the rows or columns in the diagram below are not complete, they are not examples of arrays:

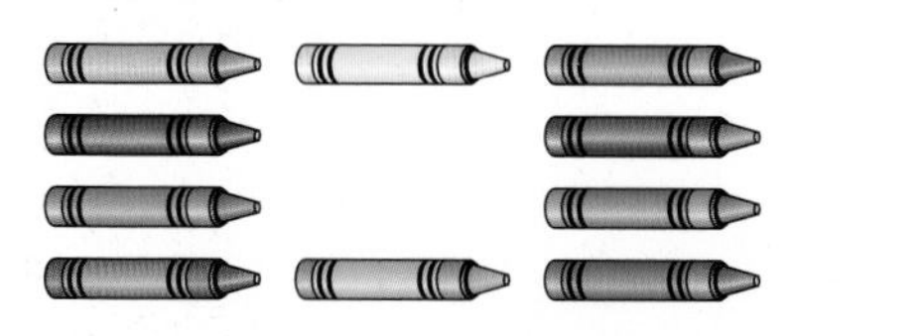

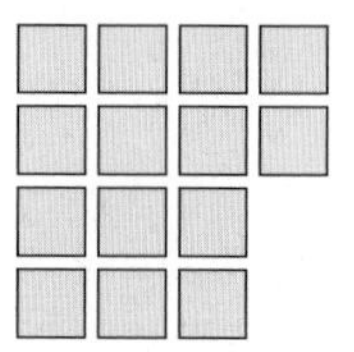

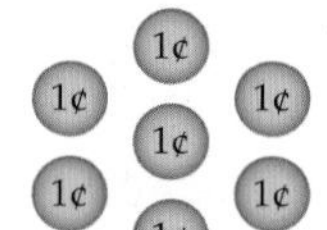

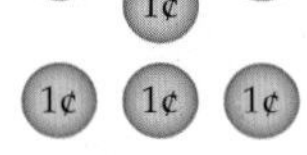

Arrays are usually described by saying the number of rows first, then the number of columns.

Example

This diagram shows a 4-by-5 array. That means it has 4 rows and 5 columns.

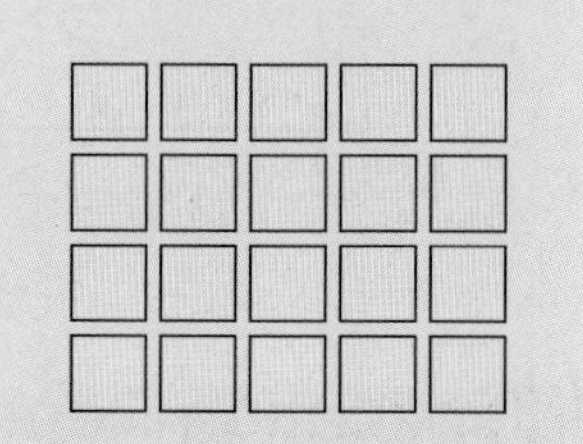

It could represent the total number of pencils in a situation in which four students (rows) each have pencils in five different colors (columns).

If you count by rows, this array could represent the total number of pencils by adding the number of pencils that each students has.

$5 + 5 + 5 + 5 = 20$ pencils is represented by the twenty squares in the array.

If you count by columns, this array could represent the total number of pencils by adding the number of pencils in each color:

$$4 + 4 + 4 + 4 + 4 = 20$$

Arrays show multiplication. This array shows either four groups of five, which is $4 \bullet 5 = 20$, or five groups of four, which is $5 \bullet 4 = 20$.

This array also shows $4 \bullet 5 = 5 \bullet 4$. This is an example of the commutative property of multiplication: $a \bullet b = b \bullet a$.

Multiplication Represented in Areas

The area of rectangles can be used to represent multiplication of whole numbers, decimals, and fractions.

Example

This rectangle represents multiplication of decimals:

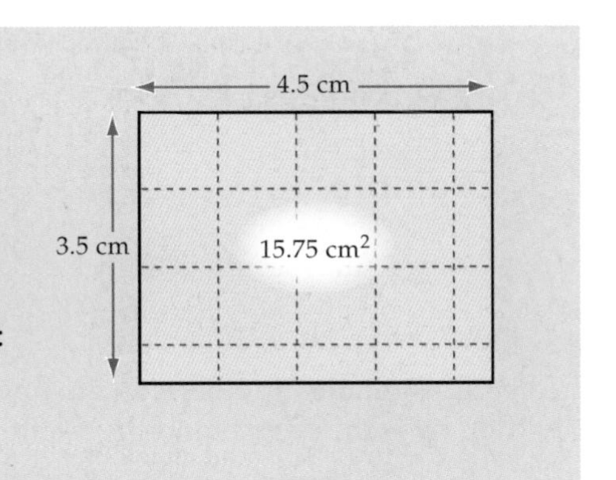

The rectangle has sides of length 4.5 cm and 3.5 cm.

The area of the rectangle is the product of the lengths of its sides: 4.5 cm • 3.5 cm = 15.75 cm^2.

4.5 • 3.5 = 15.75 cm^2 or
3.5 • 4.5 = 15.75 cm^2

You can also use the distributive property to multiply these two numbers.

Example

$4.5 \bullet 3.5$
$= (4 + 0.5) \bullet (3 + 0.5)$
$= (4 \bullet 3) + (4 \bullet 0.5) + (0.5 \bullet 3) + (0.5 \bullet 0.5)$
$= 12 + 2 + 1.5 + 0.25$
$= 15.75$ cm^2

Notice how this rectangle shows the sum as four separate areas: 12 + 2 + 1.5 + 0.25

4 cm
0.5 cm
3 cm
0.5 cm
12 cm^2
1.5 cm^2
2 cm^2
0.25 cm^2

Total Area =15.75 cm^2

The Concept of Division

Division can be shown in four ways:

- Using the symbol "÷", as in $20 \div 5 = 4$
- As a fraction, as in $\frac{20}{5} = 4$
- Using $\overline{)}$, as in $5\overline{)20}$ with quotient 4
- Sometimes using the symbol "/", as in $20/5 = 4$

Example

$63 \div 9$ $\qquad 9\overline{)63}$ $\qquad \frac{63}{9}$ $\qquad 63/9$

In any division, the number you are dividing into is the *dividend*, the number you are dividing by is the *divisor*, and the result is the *quotient*.

Example

$63 \div 9 = 7$

63 is the dividend, 9 is the divisor, and 7 is the quotient.

In general, for numbers other than zero:

If you know a multiplication fact, then you also know two division facts.

If $a \bullet b = c$, (and none of these numbers is zero) then $c \div b = a$ and $c \div a = b$.

Example

If $3 \bullet 4 = 12$, then $12 \div 3 = 4$ and $12 \div 4 = 3$.

The division $12 \div 4 = 3$ can be thought of as a short representation of the repeated subtraction $12 - 4 - 4 - 4 = 0$. Subtracting the number 4 from 12 a total of 3 times equals 0. This tells us that $12 \div 4 = 3$.

Division can be thought of as finding the number of columns that are used when a given number of items are organized into a given number of rows.

Example

Here are the ways to organize 12 items into rows and columns.

If there is 1 row, there must be 12 columns. $12 \div 1 = 12$	If there are 2 rows, there must be 6 columns. $12 \div 2 = 6$	If there are 3 rows, there must be 4 columns. $12 \div 3 = 4$

If there are 12 rows, there must be 1 column. $12 \div 12 = 1$	If there are 6 rows, there must be 2 columns. $12 \div 6 = 2$	If there are 4 rows, there must be 3 columns. $12 \div 4 = 3$

If you know the product is 12 and one of the numbers is 3, then you have $3 \bullet \blacksquare = 12$. The other number must be 4 because $3 \bullet 4 = 12$.

$12 \div 3 = 4$

Division Represented on the Number Line

You can represent division on a number line.

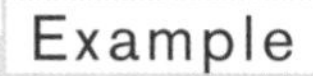

$5.6 \div 4 = 1.4$ can be represented as follows.

Start with a segment of a number line of length 5.6

Divide the segment into four equal parts

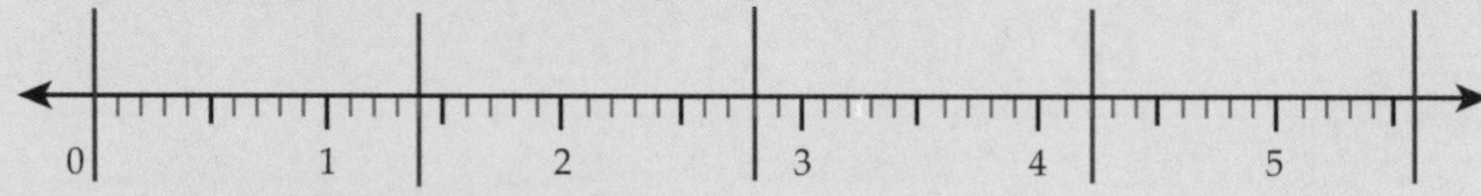

The quotient is the length of one of these equal parts, in this case 1.4

Division Represented in Arrays and Areas

Division can be represented as the process of organizing a collection of items (the dividend) into an array with a given number of rows (the divisor).

Example

$24 \div 4$

This diagram shows the start of an attempt to organize twenty-four items into four rows.

This diagram shows the completed array. It takes six columns to arrange 24 items into 4 rows.

The quotient is given by the number of columns, 6.
$24 \div 4 = 6$

Rectangles can also be used to represent division involving decimals and fractions.

Example

This diagram represents either of these two division equations:

$15.75 \text{ cm}^2 \div 4.5 \text{ cm} = 3.5 \text{ cm}$

$15.75 \text{ cm}^2 \div 3.5 \text{ cm} = 4.5 \text{ cm}$

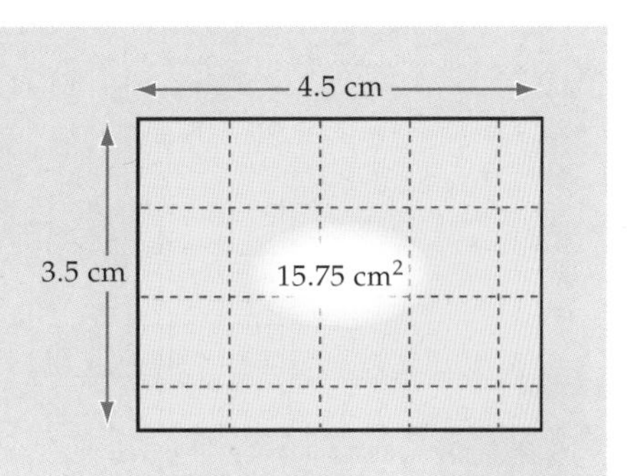

Multiplication and Division Are Inverse Operations

One way to check the result of a division calculation such as $36 \div 4$ is to multiply the result by 4. The product should be 36, the number you started with.

Example

$36 \div 4 = 9$	The result is 9.
$9 \bullet 4 = 36$	Multiply the result by the divisor.

The processes of dividing by 4 and multiplying by 4 "undo" each other. Multiplying and dividing are *inverses* of each other.

Example

This number line represents either $4 \bullet 1.4 = 5.6$ or $5.6 \div 4 = 1.4$.

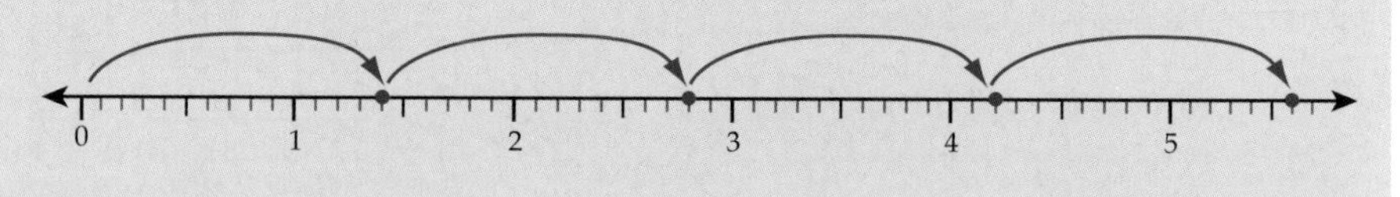

Example

These arrays represent all of the following:

$4 \bullet 5 = 20$
$20 \div 5 = 4$

$5 \bullet 4 = 20$
$20 \div 4 = 5$

The number facts $4 \bullet 5 = 20$ and $5 \bullet 4 = 20$ provide the answer to questions such as these:

- How many items are there in a 4-by-5 array?
- How many pencils are there altogether if 5 students have 4 pencils each?

The number fact $20 \div 5 = 4$ provides the answer to questions such as these:

- How many columns are needed for 20 items to fit into an array with 5 rows?
- If there are 20 pencils, how many students can be given 5 pencils each?

The number fact $20 \div 4 = 5$ provides the answer to questions such as these:

- How many columns are needed for 20 items to fit into an array with 4 rows?
- If 20 pencils are shared among 4 students, how many pencils does each student have?

In general, for numbers other than zero:

If you know a number fact using multiplication, then you also know three other number facts.

If $a \bullet b = c$, then $b \bullet a = c$, $c \div b = a$, and $c \div a = b$
for all numbers a, b, and c, with a and b not equal to 0.

If you know a number fact using division, then you also know two number facts using multiplication and another number fact using division.

If $c \div a = b$, then a $\bullet\ b = c$, $b \bullet a = c$, and $c \div b = a$
for all numbers a, b, and c, with a and b not equal to 0.

Basic Facts for Multiplication

For many multiplication problems, you can use your knowledge of place value and the number facts to figure out answers in your head.

You need to know all the multiplication facts for all pairs of single-digit numbers. With this knowledge you can perform more difficult calculations quickly and efficiently, without relying on a calculator. Here are the multiplication facts for 1 through 9.

Multiplication Facts

1 • 0 = 0	1 • 5 = 5	2 • 0 = 0	2 • 5 = 10	3 • 0 = 0	3 • 5 = 15
1 • 1 = 1	1 • 6 = 6	2 • 1 = 2	2 • 6 = 12	3 • 1 = 3	3 • 6 = 18
1 • 2 = 2	1 • 7 = 7	2 • 2 = 4	2 • 7 = 14	3 • 2 = 6	3 • 7 = 21
1 • 3 = 3	1 • 8 = 8	2 • 3 = 6	2 • 8 = 16	3 • 3 = 9	3 • 8 = 24
1 • 4 = 4	1 • 9 = 9	2 • 4 = 8	2 • 9 = 18	3 • 4 = 12	3 • 9 = 27
4 • 0 = 0	4 • 5 = 20	5 • 0 = 0	5 • 5 = 25	6 • 0 = 0	6 • 5 = 30
4 • 1 = 4	4 • 6 = 24	5 • 1 = 5	5 • 6 = 30	6 • 1 = 6	6 • 6 = 36
4 • 2 = 8	4 • 7 = 28	5 • 2 = 10	5 • 7 = 35	6 • 2 = 12	6 • 7 = 42
4 • 3 = 12	4 • 8 = 32	5 • 3 = 15	5 • 8 = 40	6 • 3 = 18	6 • 8 = 48
4 • 4 = 16	4 • 9 = 36	5 • 4 = 20	5 • 9 = 45	6 • 4 = 24	6 • 9 = 54
7 • 0 = 0	7 • 5 = 35	8 • 0 = 0	8 • 5 = 40	9 • 0 = 0	9 • 5 = 45
7 • 1 = 7	7 • 6 = 42	8 • 1 = 8	8 • 6 = 48	9 • 1 = 9	9 • 6 = 54
7 • 2 = 14	7 • 7 = 49	8 • 2 = 16	8 • 7 = 56	9 • 2 = 18	9 • 7 = 63
7 • 3 = 21	7 • 8 = 56	8 • 3 = 24	8 • 8 = 64	9 • 3 = 27	9 • 8 = 72
7 • 4 = 28	7 • 9 = 63	8 • 4 = 32	8 • 9 = 72	9 • 4 = 36	9 • 9 = 81

Mental Strategies for Multiplication

- Any number multiplied by zero is zero. This is the zero property of multiplication.

Example

$4 \bullet 0 = 0$ $18 \bullet 0 = 0$ $29 \bullet 0 = 0$ $48 \bullet 0 = 0$

- Any number multiplied by one remains unchanged. This is the identity property of multiplication.

Example

$8 \bullet 1 = 8$ $23 \bullet 1 = 23$ $32 \bullet 1 = 32$ $55 \bullet 1 = 55$

- Any number multiplied by a power of ten moves each digit to the left or right. The number of places moved is determined by the number of zero digits in the powers of ten.

Example

$48 \bullet 100 = 4{,}800$

Thousands	Hundreds	Tens	Ones
		4	8
4	8	0	0

Each digit is moved two place to the left, because there are two zeros in 100. The result is 4,800.

Chapter 7

Multiplication of Multidigit Numbers

You can use arrays to show multiplication of multidigit numbers.

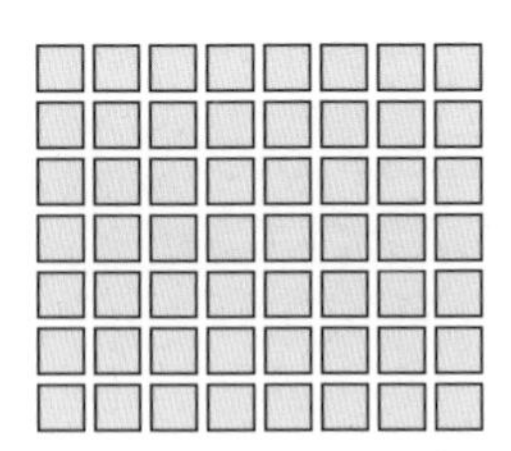

Look at the 7-by-8 array on the right.

Because it is easier to remember multiplication facts with smaller numbers, you can "break," or "decompose" the large array into smaller arrays.

This can be done in several different ways. Two examples are shown below.

Example

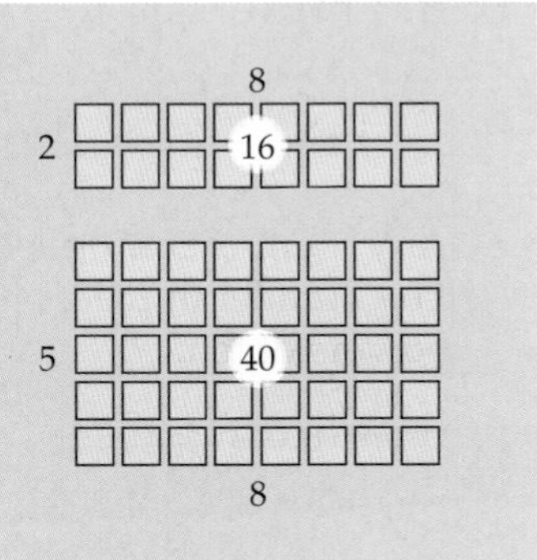

Since 2 + 5 = 7, you could make a 2-by-8 array and a 5-by-8 array.

If you remember that 2 • 8 is 16 and that 5 • 8 is 40, you can find 7 • 8 easily. You can add the number of items in the two smaller arrays:

$$2 \bullet 8 + 5 \bullet 8 = 16 + 40 = 56$$
$$7 \bullet 8 = 56$$

Example

Here is a way to break the 7-by-8 array into 4 smaller arrays.

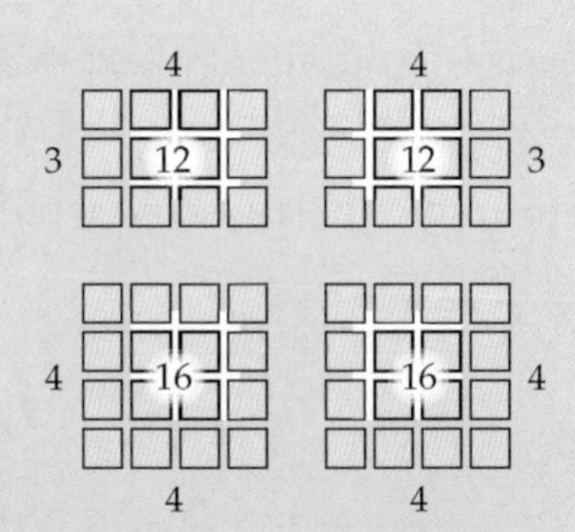

This way of breaking the 7-by-8 array uses the addition facts, 3 + 4 = 7 and 4 + 4 = 8, and the multiplication facts, 3 • 4 = 12 and 4 • 4 = 16.

Combined, the four small arrays make the big array.

The product of 7 and 8 is the sum of the partial products:

12 + 12 + 16 + 16 = 56

You can use the same strategies with multidigit numbers.

Example

23 • 34 =

To multiply 23 • 34 break it down into a 23-by-30 array and a 23-by-4 array.

You could break down the 23 by 30 further into a 20 by 30 and a 3 by 30. You could also do the same thing with the 23 by 4.

That would give you:

20 by 30 = 600
3 by 30 = 90
20 by 4 = 80
3 by 4 = 12
or 600 + 90 + 80 + 12 = 782

Place Value in Multiplication and Division

Our base-10 number system makes it easy to multiply or divide any number by 10, 100, 1,000, and so on.

The examples below show what happens when a number is multiplied by 10.

Example

$$10 \bullet \frac{1}{1000} = \frac{1}{100}$$

Each thousandth becomes a hundredth. The digit moves from the thousandths place to the hundredths place:

$$10 \bullet 0.001 = 0.01$$

Example

$$10 \bullet \frac{1}{100} = \frac{1}{10}$$

Each hundredth becomes a tenth. The digit moves from the hundredths place to the tenths place:

$$10 \bullet 0.01 = 0.1$$

Example

$$10 \bullet \frac{1}{10} = 1$$

Each tenth becomes a one. The digit moves from the tenths place to the ones place, on the other side of the decimal point:

$$10 \bullet 0.1 = 1$$

When a number is multiplied by 10, the decimal point moves to the right one place.

Example

$27.54 \bullet 10 = 275.4$ $\qquad$ $2{,}754 \bullet 10 = 27{,}540$

Dividing by 10 is the inverse of multiplying by 10. When a number is divided by 10 it decreases, and the decimal point moves to the left one place.

Example

$2{,}754 \div 10 = 275.4$ $\qquad$ $0.27 \div 10 = 0.027$

The Partial Products Method for Multiplication

The partial products method for multiplication is an application of the distributive property: $a(b + c) = (a \bullet b) + (a \bullet c)$.

Example

$25 \bullet 32 = 25 \bullet (30 + 2)$ — 32 in expanded form

$= (25 \bullet 30) + (25 \bullet 2)$ — distributive property

$= (20 + 5) \bullet 30 + (20 + 5) \bullet 2$ — 25 in expanded form

$= [30 \bullet (20 + 5)] + [2 \bullet (20 + 5)]$ — commutative property

$= (30 \bullet 20) + (30 \bullet 5) + (2 \bullet 20) + (2 \bullet 5)$ — distributive property

$25 \bullet 32 = 600 + 150 + 40 + 10$

$25 \bullet 32 = 800$

The four partial products in the example below are calculated separately and then added together. This process can be represented by calculating the area of a rectangle by first breaking its sides into parts.

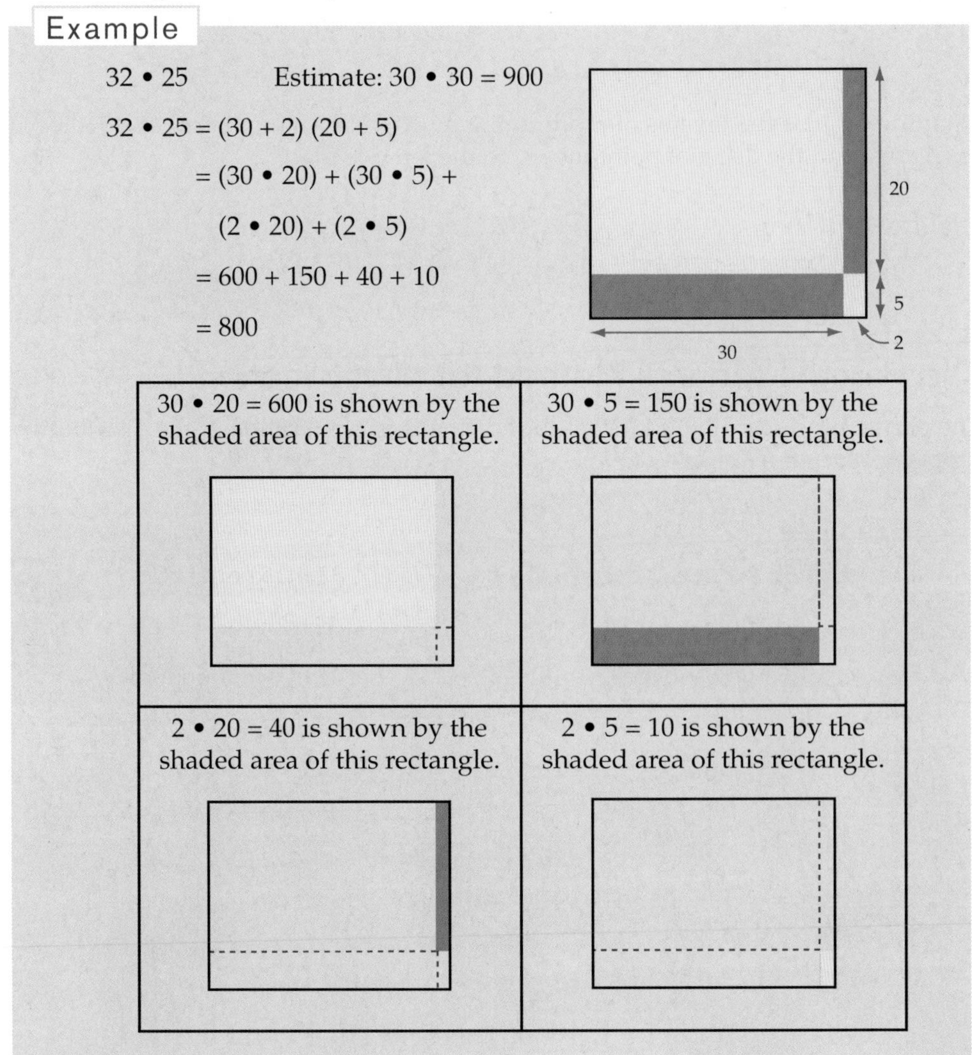

The Standard Method for Multiplication

The standard method for multiplying two multidigit numbers is sometimes called the *long multiplication algorithm*. The method involves using the distributive property, but this time with just one of the two numbers written in expanded form.

Here is 145 • 73 solved using the distributive property and by expanding 73 to 70 + 3.

Example

145 • 73	Estimate: 100 • 100 = 10,000
145 • 73 = 145 • (70 + 3)	73 written in expanded form
= (145 • 70) + (145 • 3)	distributive property
= 10,150 + 435	
= 10,585	

Here is 145 • 73 solved using an area model by expanding 73 to 70 + 3.

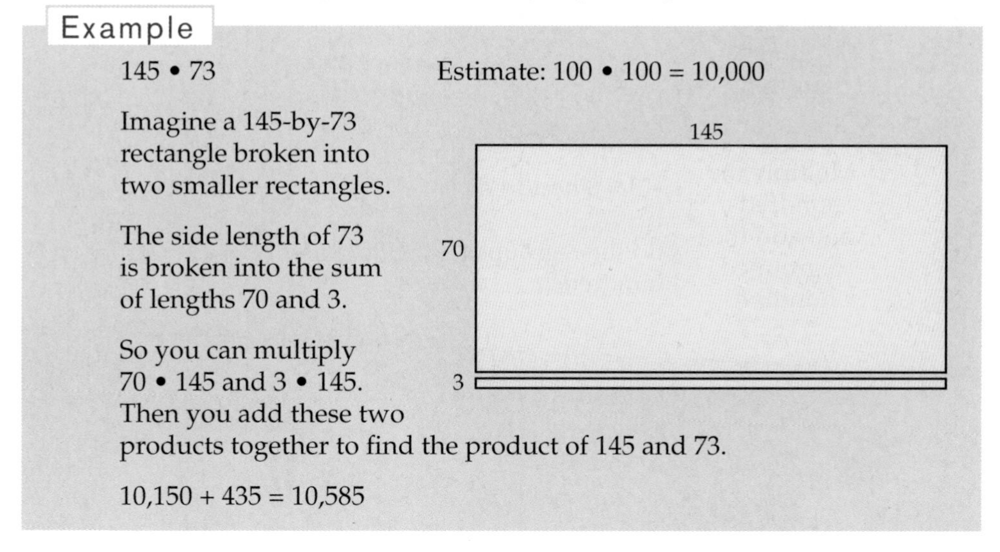

Example

145 • 73 Estimate: 100 • 100 = 10,000

Imagine a 145-by-73 rectangle broken into two smaller rectangles.

The side length of 73 is broken into the sum of lengths 70 and 3.

So you can multiply 70 • 145 and 3 • 145. Then you add these two products together to find the product of 145 and 73.

10,150 + 435 = 10,585

Here is 145 • 73 solved using the standard multiplication method.

Example

145 • 73 Estimate: 100 • 100 = 10,000

Step 1: Multiply 145 • 3

Multiply the ones place. Regroup into tens.	5 • 3 = 15 ones 15 ones = 1 ten + 5 ones Write the 5 in the ones place and move the 1 into the tens place.	Hundreds / Tens / Ones (regrouped 1 above tens) 1 4 5 × 7 3 5
Multiply the tens place and add the regrouped ten. Regroup into hundreds.	4 • 3 = 12 tens Add the regrouped 1 to 12 to make 13 tens. 13 tens = 10 tens + 3 tens = 1 hundred + 3 tens Write the 3 in the tens place and move the 1 into the hundreds place.	Hundreds / Tens / Ones (regrouped 1 above hundreds, 1 above tens) 1 4 5 × 7 3 3 5
Multiply the hundreds, and add the regrouped hundred.	1 • 3 = 3 hundreds Add the regrouped 1 to 3 to make 4 hundreds.	Hundreds / Tens / Ones (regrouped 1 above hundreds, 1 above tens) 1 4 5 × 7 3 4 3 5

(continued)

Example

145 • 73 Estimate: 100 • 100 = 10,000

Step 2: Multiply 145 • 70

Start in the ones place.	You are now multiplying by 7 tens.	Hundreds Tens Ones
Multiply 70 times the ones.	7 • 5 = 35 tens or 70 • 5 = 350 35 tens = 3 hundreds and 5 tens Write 5 under the 3 in the tens column. Regroup the 3 into the tens place.	(regrouped 3 above the 4) 1 4 5 × 7 3 4 3 5 5
Multiply the tens place.	7 • 4 = 28 hundreds or 70 • 40 = 2,800 Add the regrouped 3 to 28 to make 31 hundreds. 31 hundreds = 3 thousands and 1 hundred. Write 1 under the 4 in the hundreds column. Regroup the 3 into the hundreds place.	Hundreds Tens Ones (regrouped 3 above the 1, 3 above the 4) 1 4 5 × 7 3 4 3 5 1 5
Regroup into hundred.	7 • 1 = 7 thousands or 70 • 100 = 7,000 Add the regrouped 3 to 7 to make 10 thousands. 10 thousands = 1 ten thousand and no thousands. Write 0 in the thousands place and 1 in the ten thousands place.	Hundreds Tens Ones (regrouped 3 above the 1, 3 above the 4) 1 4 5 × 7 3 4 3 5 1 0 1 5 1 0 5 8 5
Add the digits.	Write as an equation: 145 • 73 = 10,585	

Using the standard method is similar to using the distributive property. By using expanded form, you simply use a shortcut notation to find the product.

Example

When you use the standard method for multiplying 13 • 354, you first line up the digits.

Hundreds	Tens	Ones
3	5	4
x	1	3

Next, you multiply from right to left: multiply 6 pairs of numbers to get 6 products.
Then, add the products:

$$3(4 + 50 + 300) + 10(4 + 50 + 300)$$
$$(3 \bullet 4) + (3 \bullet 50) + (3 \bullet 300) + (10 \bullet 4) + (10 \bullet 50) + (10 \bullet 300)$$

When you use the standard method, these products are disguised by the regrouping and by lining up the place values.

You can use the number properties to show that the standard method's sum of products $(3 \bullet 4) + (3 \bullet 50) + (3 \bullet 300) + (10 \bullet 4) + (10 \bullet 50) + (10 \bullet 300)$ is equivalent to $(3 + 10)(4 + 50 + 300)$.

Action	Justification	Comment
$(3 \bullet 4) + (3 \bullet 50) + (3 \bullet 300) + (10 \bullet 4) + (10 \bullet 50) + (10 \bullet 300)$	Standard method	Given
$3(4 + 50 + 300) + (10 \bullet 4) + (10 \bullet 50) + (10 \bullet 30)$	$a \bullet b + a \bullet c = a(b + c)$ Distributive property	Put 3 in the *a* position of $a(b + c)$
$3(4 + 50 + 300) + 10(4 + 50 + 300)$	$a \bullet b + a \bullet c = a(b + c)$ Distributive property	Put 10 in the *a* position of $a(b + c)$
$(4 + 50 + 300)3 + (4 + 50 + 300)10$	$a \bullet b = b \bullet a$ Commutative property	Put $(4 + 50 + 300)$ in the *a* position of $b \bullet a$
$(4 + 50 + 300)(3 + 10)$	$a \bullet b + a \bullet c = a(b + c)$ Distributive property	$(4 + 50 + 300)$ is *a*, 3 is *b*, and 10 is *c*
$(3 + 10)(4 + 50 + 300)$	$a \bullet b = b \bullet a$ Commutative property	

Partial Quotients Method for Division

Just as multiplication can be done using repeated addition, division can be done using repeated subtraction of multiples of the divisor from the dividend.

Example

$2,820 \div 16$ Estimate: $3,000 \div 20 = 150$

Start by listing some easy multiples of 16.
$100 \bullet 16 = 1,600$ $50 \bullet 16 = 800$ $25 \bullet 16 = 400$ $1 \bullet 16 = 16$

Now start subtracting multiples of 16 from 2,820. Continue until the remainder is less than 16, the divisor.

Subtract $100 \bullet 16 = 1,600$ from 2,820, giving 1,220.
The 100 will be part of the quotient.

Subtract $50 \bullet 16 = 800$ from 1,220, giving 420.
The 50 will be part of the quotient.

Subtract $25 \bullet 16 = 800 \div 2 = 400$ from 420, giving 20.
The 25 will be part of the quotient.

Subtract $1 \bullet 16 = 16$ from 20, giving 4.
The 1 will be part of the quotient.

16)2820	Partial Quotients
−1600	100
1220	
−800	50
420	
−400	25
20	
−16	1
4	

The remainder, 4, is less than 16 and so the process stops. The sum of the partial quotients is $100 + 50 + 25 + 1 = 176$, Remainder 4.

Result: $2,820 \div 16 = 176\frac{4}{16} = 176\frac{1}{4}$

Alternatively, continue by subtracting decimal fractions of 16.

$0.2 \bullet 16 = 32$ tenths = 3 ones and 2 tenths
$0.05 \bullet 16 = 80$ hundredths = 8 tenths

4.0	
−3.2	0.2
0.8	
−0.8	0.05
0	

Add these up: $176 + 0.2 + 0.05 = 176.25$
Write as an equation: $2,820 \div 16 = 176.25$

The Standard Method for Division

The partial quotients method involves a lot of writing. A quicker method, called the *short division method*, can be used when the divisor is a single-digit.

Example

Divide 36 by 4.

The division is $36 \div 4 = \square$
which can be rewritten as $4 \bullet \square = 36$.

What is the Method?	What is Really Happening?
Look at the left-most digit of the dividend. Ask: $4 \bullet \square = 36$? There is no whole number that works.	Regroup the 3 tens as 30 ones. 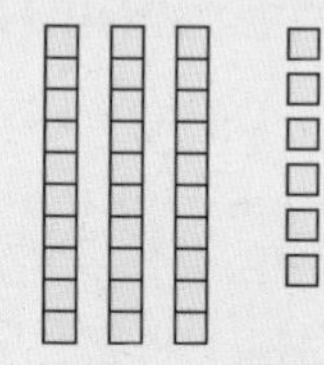
Write the 3 tens by the 6 ones to make 36 ones. $\begin{array}{r} 9 \\ 4\overline{)3^{3}6} \end{array}$ Say: $4 \bullet \square = 36$. The result is 9.	Now there are 36 ones. Make 9 groups of 4. 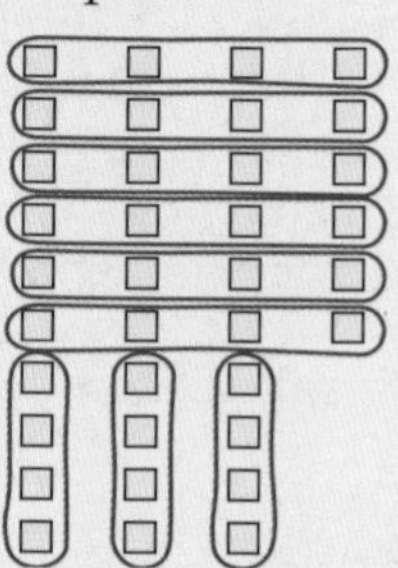

Write as an equation: $36 \div 4 = 9$

Example

139 ÷ 6

The product is 6 • ▮ = 139.

What is the Method?	Write This.	What is Really Happening?
Look at the hundreds digit and ask: 6 • ▮ = 1? There is not a whole number that works.	$6\overline{)1\ 3\ 9}$	
Write the 1 hundred by the 3 tens to make 13 tens. Look at the tens and ask: 6 • ▮ = 13? 6 • 2 = 12 is closest to 13.	$6\overline{)1\,^{1}3\ 9}$	Regroup the 1 hundred as 10 tens. There are 13 tens. The result so far is 20, 2 is in the tens place.
Write the 2 in the tens column of the quotient. Write the 1 ten by the 9 to make 19 ones.	$\begin{array}{r} 2 \\ 6\overline{)1\,^{1}3\,^{1}9} \end{array}$	6 • 20 = 120 Regroup the 1 ten for 10 ones. There are 19 ones. The remainder is 139 – 120 = 19.
Look at the ones and ask: 6 • ▮ = 19? 6 • 3 = 18 is closest to 19. Write a 3 in the ones column of the quotient.	$\begin{array}{r} 2\ 3 \text{ R1} \\ 6\overline{)1\,^{1}3\,^{1}9} \end{array}$	You can make 23 groups of 6 with 1 remaining.
Stop here, with 139 ÷ 6 = 23 Remainder 1.		

Write as an equation: 139 ÷ 6 = 23 Remainder 1

Example

$139 \div 6 = 23$ Remainder 1

You can also write the remainder as a decimal.

What is the Algorithm?	Write This.	What is Really Happening?
To write the quotient as a decimal, regroup the 1 into the tenths place to make 10 tenths. Ask: $6 \bullet \square = 10$	$\begin{array}{r} 23. \\ 6\overline{)1{}^{1}3{}^{1}9.{}^{1}000} \end{array}$	Regroup one unit as 10 tenths.
The result is 1 (in the tenths place), with 4 remaining. Write the 4 tenths by the hundredths place to make 40 hundredths.	$\begin{array}{r} 23.1 \\ 6\overline{)1{}^{1}3{}^{1}9.{}^{1}0{}^{4}00} \end{array}$	Regroup 4 tenths as 40 hundredths.
Look at the hundredths place and ask: $6 \bullet \square = 40$? The result is 6, with 4 remaining.	$\begin{array}{r} 23.16 \\ 6\overline{)1{}^{1}3{}^{1}9.{}^{1}0{}^{4}0{}^{4}0} \end{array}$	$6 \bullet 0.6 = 0.36$ Regroup the 4 hundredths as 40 thousandths.
Continuing in this way gives the result, a repeating decimal.	$\begin{array}{r} 23.166 \\ 6\overline{)1{}^{1}3{}^{1}9.{}^{1}0{}^{4}0{}^{4}0{}^{4}0} \end{array}$	

Write as an equation: $139 \div 6 = 23.166666\ldots$

Multiplying and Dividing Decimals

Multiplying by a Decimal

Example

14.5 • 0.73

Start by temporarily ignoring the decimal points so that you can calculate with whole numbers. Use 145 instead of 14.5; use 73 instead of 0.73.

145 • 73 = 10,585

Then use rounded numbers to decide where to place the decimal point. Use 15 instead of 14.5; use 0.7 instead of 0.73.

$$15 \bullet 0.7 = 15 \bullet \frac{7}{10} = 15 \bullet 7 \bullet \frac{1}{10} = 105 \bullet \frac{1}{10} = 10.5 \text{ (or } 10.500\text{)}$$

Look at your decimal estimate and the whole number answer:

Estimated: 10.500 Whole number: 10,585

Use the whole-number digits to write the decimal result.

10.585

14.5 • 0.73 = 10.585

Dividing by a Decimal

Example

28.4 ÷ 1.6

Use the same steps as in multiplying.

First, calculate using whole numbers.

284 ÷ 16 = 17.75

The digits of the result are 1,775.

Then estimate the decimal value. Round the numbers. Use 30 instead of 28.4 and 2 instead of 1.6.

30 ÷ 2 = 15

Look at your estimate and the digits of the result:

Estimated: 15.00 Digits of the result: 1,775

Use the digits of the result to write the decimal result.

17.75
28.4 ÷ 1.6 = 17.75

FACTORS AND MULTIPLES

CHAPTER 8

Number Sets

The numbers used for counting are called the *natural numbers*. The set of natural numbers is:

$$N = \{1, 2, 3, 4, 5, 6, 7, 8, 9, 10, 11, 12, \ldots\}$$

On the number line, the natural numbers can be represented like this:

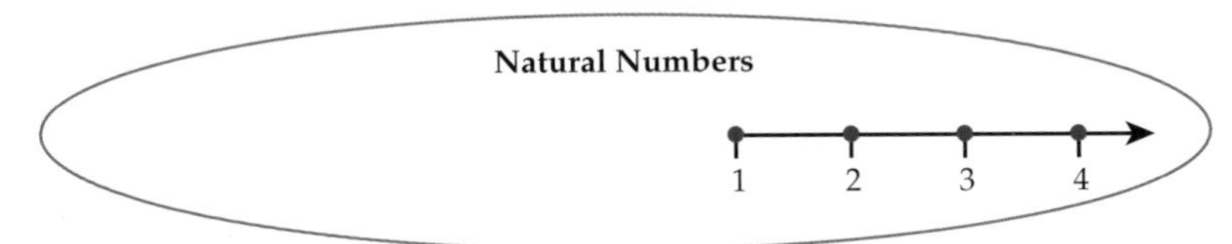

The number zero is not a natural number, but it does belong to the set of *whole numbers*. The set of whole numbers is:

$$W = \{0, 1, 2, 3, 4, 5, 6, 7, 8, 9, 10, 11, 12, \ldots\}$$

On the number line, the whole numbers can be represented like this:

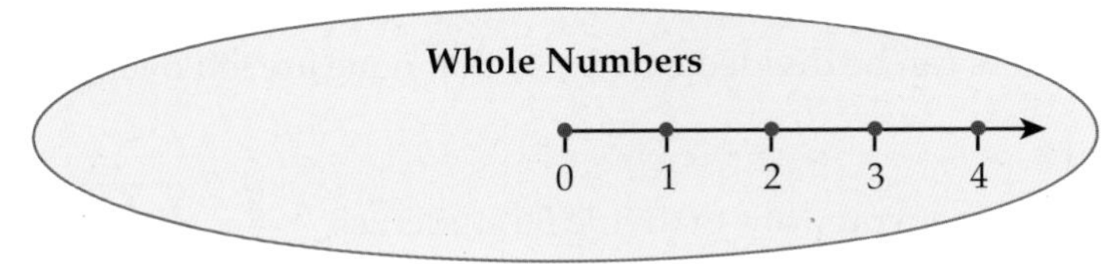

Integers include the following:

- The positive whole numbers
- Their opposites—the negative whole numbers
- Zero (zero is neither negative nor positive)

The set of integers is:

$$I = \{\ldots, -4, -3, -2, -1, 0, 1, 2, 3, 4, \ldots\}$$

On the number line, the integers can be represented like this:

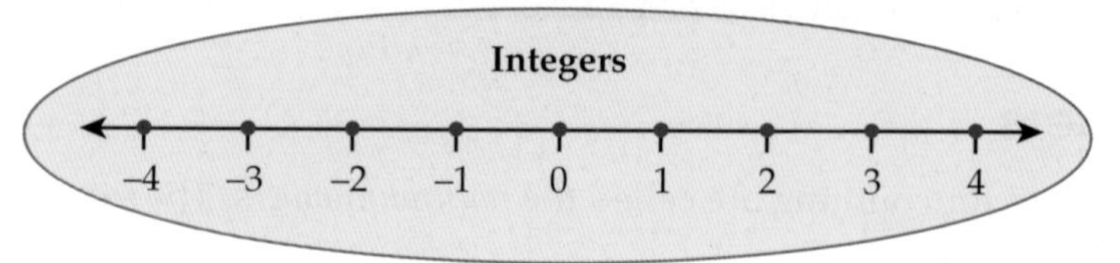

In summary:

- *Natural numbers* are all the numbers you use to count.
  (1, 2, 3, 4,…)
- *Whole numbers* include all of the natural numbers and zero.
  (0, 1, 2, 3, 4,…)
- *Integers* include the natural numbers, their opposites, and zero.
  (…, –4, –3, –2, –1, 0, 1, 2, 3, 4,…)

The integers are sometimes organized this way:

the positive integers	(1, 2, 3, 4,…)
the negative integers	(…, –4, –3, –2, –1)
zero	(neither positive nor negative)

Remainders

Sometimes, numbers can be divided evenly, so that none are left over.

Example

Imagine that you want to divide 36 students into groups of 9 students.
How many groups can be made?

You can make 4 groups of 9 students.

Here is how to write this with an equation using division: $36 \div 9 = 4$.

The division equation in the example is $36 \div 9 = 4$.

36 is the *dividend*, 9 is the *divisor*, and 4 is the *quotient*.

Notice that nothing is left over when you divide a group of 36 into groups of 9. You can say there is "no remainder" when 36 is divided by 9.

There are several ways of writing a division problem.

Example

Here are some ways to say $36 \div 9$:

36 divided by 9

36 is divisible by 9

9 is a divisor of 36

9 divides 36

9 goes into 36 evenly

Sometimes, division will result in a remainder.

Example

Imagine you want to divide 36 students into groups.

If you try to divide 36 students into groups of 7 students, you get 5 groups of 7 students.

This time, one student of the 36 remains after the 5 groups of 7 have been formed.

Here is how to write the equation using division:

$36 \div 7 = 5$ Remainder 1.

Because there is a remainder of 1, you say that 36 is not evenly divided by 7.

When you divide one natural number by another, either there is a remainder or there is no remainder.

The remainder must always be smaller than the divisor. Otherwise, there would be enough to make one more group.

Example

If you divide natural numbers by 5, these are the possible results:

If n Equals Any of These:	The Remainder of $n \div 5$ Is:
5, 10, 15, 20, 25, 30, 35, 40,...	0
1, 6, 11, 16, 21, 26, 31, 36,...	1
2, 7, 12, 17, 22, 27, 32, 37,...	2
3, 8, 13, 18, 23, 28, 33, 38,...	3
4, 9, 14, 19, 24, 29, 34, 39,...	4

You can state facts about remainders using either division or multiplication.

For natural numbers n, k, and a:

Remainder Fact	Fact Stated Using Division	Fact Stated Using Multiplication
n is divisible by k (with no remainder)	$\frac{n}{k} = a$	$n = a \bullet k$
n is divisible by k with a remainder of r	$\frac{n}{k} = a$ (Remainder r)	$n = (a \bullet k) + r$ where $0 < r < k$

Factors and Multiples

Consider this equation: $\frac{36}{9} = 4$. Because 36 is divisible by 9 with no remainder, you say 9 is a *factor* of 36. $36 \div 4 = 9$ is also true, so 4 is a factor of 36. 9 and 4 are called a *factor pair* of 36. Can you think of other natural numbers that are factors of 36?

Consider the equivalent equation using multiplication: 9 • 4 = 36.
You say that 36 is a *multiple* of 9.

4 • 9 = 36 is also true, so 36 is also a multiple of 4. Can you think of other natural numbers that are multiples of 4 or of 9?

Now, think about this equation: $\frac{36}{7}$ = 5 Remainder 1.
Since there is a remainder, you say:

- 7 is not a factor of 36.
- 5 is not a factor of 36.
- 36 is not a multiple of 5.
- 36 is not a multiple of 7.
- 5 and 7 are not a factor pair for 36 because 5 • 7 ≠ 36.

You can also use a multiplication table to see factors and multiples.
Find the number 24 in the multiplication table.

	Columns												
Rows	×	1	2	3	4	5	6	7	8	9	10	11	12
	1	1	2	3	4	5	6	7	8	9	10	11	12
	2	2	4	6	8	10	12	14	16	18	20	22	24
	3	3	6	9	12	15	18	21	24	27	30	33	36
	4	4	8	12	16	20	24	28	32	36	40	44	48
	5	5	10	15	20	25	30	35	40	45	50	55	60
	6	6	12	18	24	30	36	42	48	54	60	66	72
	7	7	14	21	28	35	42	49	56	63	70	77	84
	8	8	16	24	32	40	48	56	64	72	80	88	96
	9	9	18	27	36	45	54	63	72	81	90	99	108
	10	10	20	30	40	50	60	70	80	90	100	110	120
	11	11	22	33	44	55	66	77	88	99	110	121	132
	12	12	24	36	48	60	72	84	96	108	120	132	144

The 24 in the second row represents 2 • 12 = 24. A class of 24 students could be divided into 2 groups with 12 in each group. You say:

24 is a multiple of 2.	2 is a factor of 24.
24 is a multiple of 12.	12 is a factor of 24.
2 and 12 are a factor pair for 24.	2 • 12 is a factorization of 24.

The 24 in the third row represents 3 • 8 = 24. A class of 24 students could be divided into 3 groups with 8 in each group. You say:

24 is a multiple of 3.	3 is a factor of 24.
24 is a multiple of 8.	8 is a factor of 24.
3 and 8 are a factor pair for 24.	3 • 8 is a factorization of 24.

The 24 in the fourth row represents 4 • 6 = 24. A class of 24 students could be divided into 4 groups with 6 in each group. You say:

24 is a multiple of 4.	4 is a factor of 24.
24 is a multiple of 6.	6 is a factor of 24.
4 and 6 are a factor pair for 24.	4 • 6 is a factorization of 24.

There is no 24 in the fifth row. Why not?

Example

If you tried to divide a class of 24 students into groups of 5 students, you would get 4 groups of 5 students, but there would be a remainder of 4 students left over.

You say:

24 is not a multiple of 5.	5 is not a factor of 24.
$24 = (5 \bullet 4) + 4$	
$24 \div 5 = 4$ Remainder 4	Here the division $24 \div 5$ is expressed using remainders.
$24 \div 5 = 4\frac{4}{5}$	Here the division $24 \div 5$ is expressed using fractions.

The multiplication table shows that 24 is a *multiple* of the numbers 2, 3, 4, 6, 8, and 12. Although the table does not show it, 24 is also a multiple of 1 and 24, since $1 \bullet 24 = 24$.

The table also shows that the *factors* of 24 are 1, 2, 3, 4, 6, 8, 12, and 24. The numbers 5, 7, 9, 10, 11, and 13 through 23 are not factors of 24.

Prime Numbers and Composite Numbers

Every natural number, n, is divisible by 1 and by itself, since $1 \bullet n = n$.

If a number greater than 1 has no factors other than 1 and itself, it is called a *prime number* (or just *prime*).

Example

The number 7 has only the factors 1 and 7; so 7 is prime.

A prime number of dots can be arranged into just one rectangular array (since the number has only 1 and itself as factors). This assumes that the array 1×7 is considered the same as the array 7×1.

Example

Seven is prime, because 7 dots can be arranged into only one rectangular array:

○ ○ ○ ○ ○ ○ ○

$1 \bullet 7$

If a number also has factors other than 1 and itself, it is not prime; it is a *composite number*.

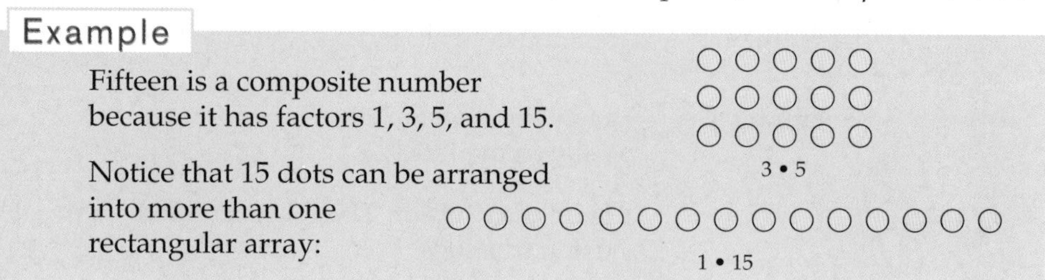

The number 1 is special. It is neither prime nor composite.
All natural numbers other than 1 are *either* prime or composite.
No number can be *both* prime and composite.

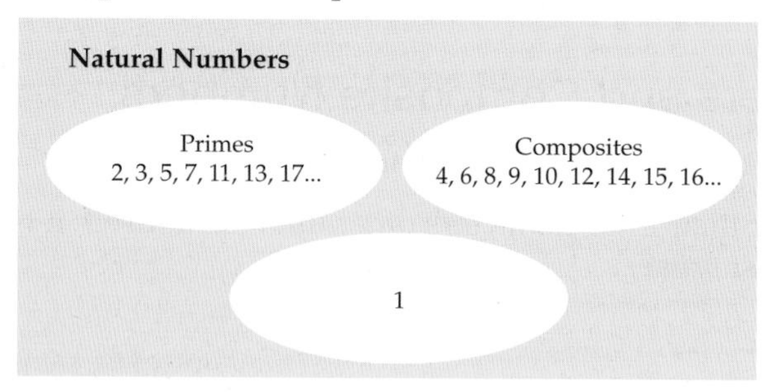

Factoring into Primes

The number 24 is a composite number. What are the factors of 24?

You can write 24 as the product of factors in several ways:

$24 = 1 \bullet 24$ $\quad 24 = 2 \bullet 2 \bullet 2 \bullet 3$ $\quad 24 = 2 \bullet 12$ $\quad 24 = 3 \bullet 8$

$24 = 2 \bullet 3 \bullet 4$ $\quad 24 = 4 \bullet 6$ $\quad 24 = 2 \bullet 2 \bullet 6$

Each of these multiplications is called a *factorization* of 24.

Look at each factorization of 24. One of them contains only *prime numbers* as *factors*. Which factorization is it?

When you factor 24 into its prime factors, you get 24 = 2 • 2 • 2 • 3.
You call this the *prime factorization* of 24.

Every natural number greater than 1 can be written as a product of prime factors.

How can you find the prime factors of a number? One way is to start with any factor pair, then break those factors into smaller factors until all the factors are primes.

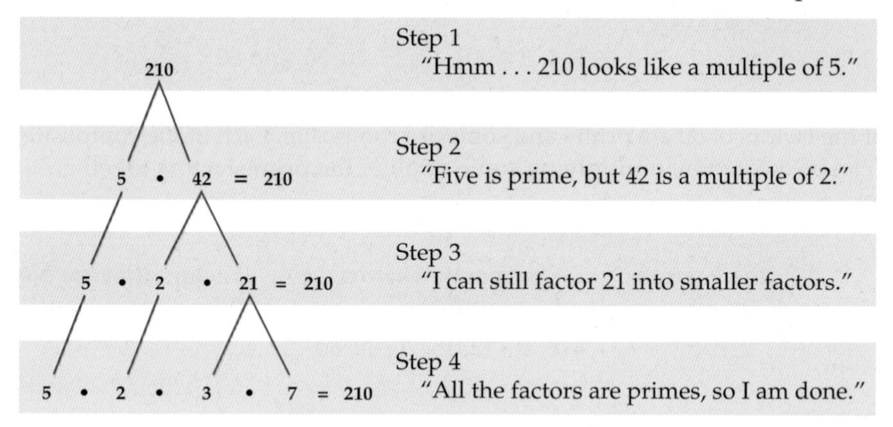

The prime factorization of 210 is 5 • 2 • 3 • 7.

Each natural number has factor pairs.

Example

Factors pairs for the number 60:

1 • 60, 2 • 30, 3 • 20, 4 • 15, 5 • 12, 6 • 10

Factors for 60:

1, 2, 3, 4, 5, 6, 10, 12, 15, 20, 30, 60

It is important not to confuse the set of factors of a number with the number's prime factorization.

Example

Prime factorization for 60:

$$60 = 2 \bullet 2 \bullet 3 \bullet 5$$

The factors of 60:

1, 2, 3, 4, 5, 6, 10, 12, 15, 20, 30, and 60

Some of the factors of 60 are prime and some are composite. Each of the composite factors can be created by multiplying some or all of the prime factors together.

Example

Prime Factors of 60	Composite Factors of 60	Factors that are Not Prime or Composite
2, 3, 5	4, 6, 10, 12, 15, 20, 30, 60	1

The Fundamental Theorem of Arithmetic

The *fundamental theorem of arithmetic* states that every natural number greater than 1 is either prime or the product of unique prime factors.

What does this mean? Each natural number has its own prime factorization, different from that of any other number.

Writing the factors in a different order does not make a different prime factorization.

Example

$24 = 2 \bullet 2 \bullet 2 \bullet 3$
$24 = 3 \bullet 2 \bullet 2 \bullet 2$

These are not different prime factorizations of 24, since they have the same prime factors.

Look again at your factorizations for the number 24. You can think of several ways to write 24 as the product of factors. But if you cross out all factorizations that contain composites, you are left with just one factorization, the prime factorization.

~~1 • 24 = 24~~ 2 • 2 • 2 • 3 = 24 ~~2 • 12 = 24~~

~~3 • 8 = 24~~ ~~2 • 3 • 4 = 24~~ ~~4 • 6 = 24~~

~~2 • 2 • 6 = 24~~

When you see any natural number, you can express the number as a product of prime factors. Either it is prime (its only factors are itself and 1) or composite (it has prime factors other than itself and 1).

There is only one prime factorization for each natural number.

There is always more than one way to write a composite number as the product of factors. But only one of these ways is a prime factorization, since primes cannot be broken down into smaller factors.

Consider these four factor trees for 24:

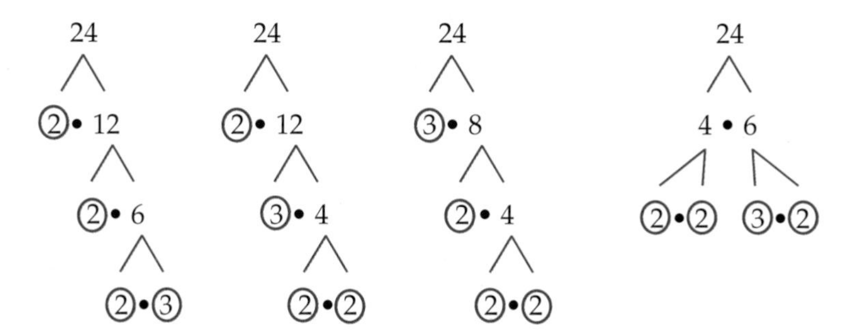

The four trees start differently, but all of them end with the same prime factorization.

The order in which you break down the composite factors into prime factors does not matter.

Any correct method leads to this unique prime factorization for the number 24:

$$2 \bullet 2 \bullet 2 \bullet 3 = 24$$

Prime factorizations can be expressed in *exponent form*.

Example

The prime factorization of 24 contains the string of factors $2 \cdot 2 \cdot 2 \cdot 3$.

In exponent form, $2 \cdot 2 \cdot 2$ is written as 2^3, where the superscript 3 represents the number of times the factor is repeated.

So, the prime factorization of 24 can be written $2^3 \cdot 3$.

Common Factors

If a number is a factor of each of two other numbers, you call it a *common factor* of those two numbers.

Example

$12 = 3 \cdot 4$ and $18 = 3 \cdot 6$
Since 3 is a factor of *both* 12 and 18, 3 is a common factor of 12 and 18.

$40 = 5 \cdot 8$ $\quad$ $48 = 8 \cdot 6$ $\quad$ $64 = 8 \cdot 8$
Since 8 is a factor of 40, 48, *and* 64, 8 is a common factor of 40, 48, and 64.

Sometimes, you want to know *all* the common factors of two or more numbers.

Example

To find all the common factors of 12 and 18:

First, list all the factors of 12: 1, 2, 3, 4, 6, 12

Then, list all the factors of 18: 1, 2, 3, 6, 9, 18

You can see that there are four numbers that appear in both lists: 1, 2, 3, 6. These are the common factors of 12 and 18.

You can show the information for common factors using a *Venn diagram*:

Example

Venn diagram for the common factors of 12 and 18:

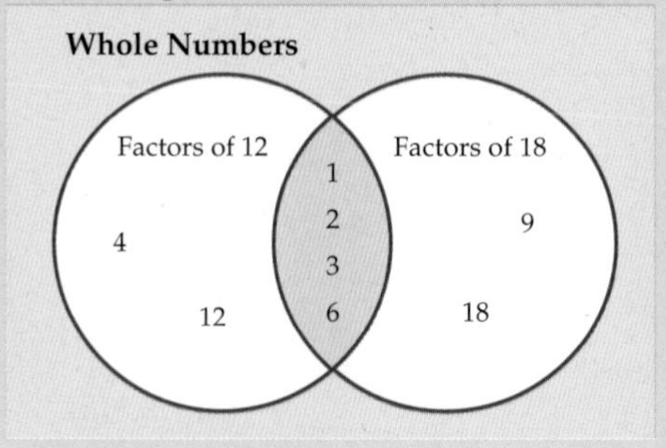

You can see the common factors of 12 and 18 where the circles overlap.

Greatest Common Factor

The greatest number that is a common factor of a group of 2 or more numbers is called the *greatest common factor*. The common factors of 12 and 18 are 1, 2, 3, and 6. Since 6 is the greatest of these, 6 is the greatest common factor of 12 and 18.

The greatest common factor is sometimes shortened to GCF.

Example

Suppose you want to find the greatest common factor of 180 and 288.

Factors of 180: 1, 2, 3, 4, 5, 6, 9, 10, 12, 15, 18, 20, 30, 36, 45, 60, 90, 180

Factors of 288: 1, 2, 3, 4, 6, 8, 9, 12, 16, 18, 24, 32, 36, 48, 72, 96, 144, 288

The greatest number that appears in both lists is 36.
This means that 36 is the greatest common factor (GCF) of 180 and 288.

Using Prime Factorization to Find the GCF

Prime factorization can help you find the greatest common factor of a pair or group of large numbers without having to list all the factors of each number.

Example

To find the greatest common factor, first find the prime factorization of 180 and 288:

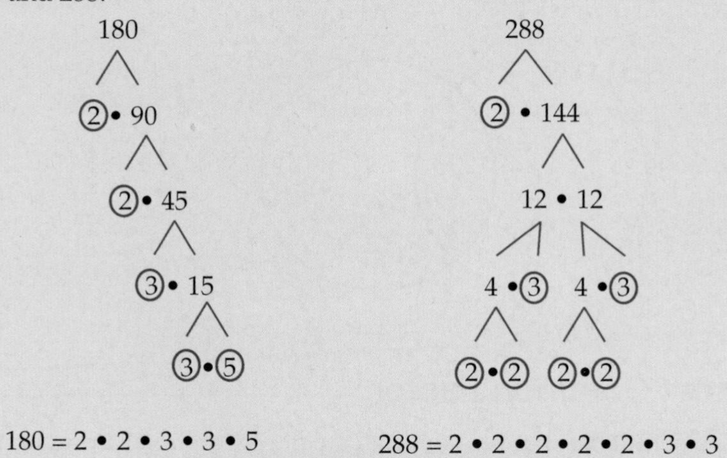

$180 = 2 \bullet 2 \bullet 3 \bullet 3 \bullet 5$ $\qquad$ $288 = 2 \bullet 2 \bullet 2 \bullet 2 \bullet 2 \bullet 3 \bullet 3$

Find the prime factors that are common to both numbers.
You take $2 \bullet 2$, since it is the longest string of 2s that is common to both factorizations.

You take $3 \bullet 3$, since it is the longest string of 3s that is common to both factorizations.

You do not take any 5s, since 5 does not occur in both factorizations.

Putting these together, you have $2 \bullet 2 \bullet 3 \bullet 3$.

The greatest common factor of 180 and 288 is equal to the product of these:

$$2 \bullet 2 \bullet 3 \bullet 3 = 36$$

This shows how 36 is a factor of both 180 and 288.

$36 \bullet 5 = 180$ $\qquad$ $36 \bullet 8 = 288$

Representing Multiples

The *multiples* of a natural number m are all numbers $n \bullet m$, where n is a natural number.

Example

Here are the multiples of the number 7:

$1 \bullet 7 = 7$

$2 \bullet 7 = 14$

$3 \bullet 7 = 21$

$4 \bullet 7 = 28$ and so on

There is an infinite number of multiples of any given natural number. Following are three different ways to represent multiples.

Multiples in a Multiplication Table

In a multiplication table, each column lists numbers that are multiples of the number at the top of the column. Also, each row lists numbers that are multiples of the number at the left end of the row.

Example

2, 4, 6, 8, 10, 12, 14,… are all multiples of 2

5, 10, 15, 20, 25, 30,… are all multiples of 5

These lists of multiples never stop. A multiplication table could go on forever, row after row, and column after column.

Here is a multiplication table for the numbers from 0 through 6.

The multiples of 3 can be found by looking at the row that begins with 3, or at the column that begins with 3.

Multiples of 3 are:

3, 6, 9, 12, 15, 18…

Columns (across) / **Rows** (down)

	0	**1**	**2**	**3**	**4**	**5**	**6**
0	0	0	0	0	0	0	0
1	0	1	2	3	4	5	6
2	0	2	4	6	8	10	12
3	0	3	6	9	12	15	18
4	0	4	8	12	16	20	24
5	0	5	10	15	20	25	30
6	0	6	12	18	24	30	36

Multiples Using Letters

The multiples of a natural number, a, are simply all numbers $a \bullet n$, where n is any natural number.

Example

The multiples of a are a, $2a$, $3a$, $4a$, $5a$,…
If $a = 4$ then the multiples are 4, 8, 12, 16, 20,…

Example

If $a = 3$ and $n = 22$, then $a \bullet n = 3 \bullet 22 = 66$.
The twenty-second number in the list of multiples of 3 is 66.

Multiples on Number Lines

Multiples can be represented as equally spaced points on the number line that starts at zero and goes on forever.

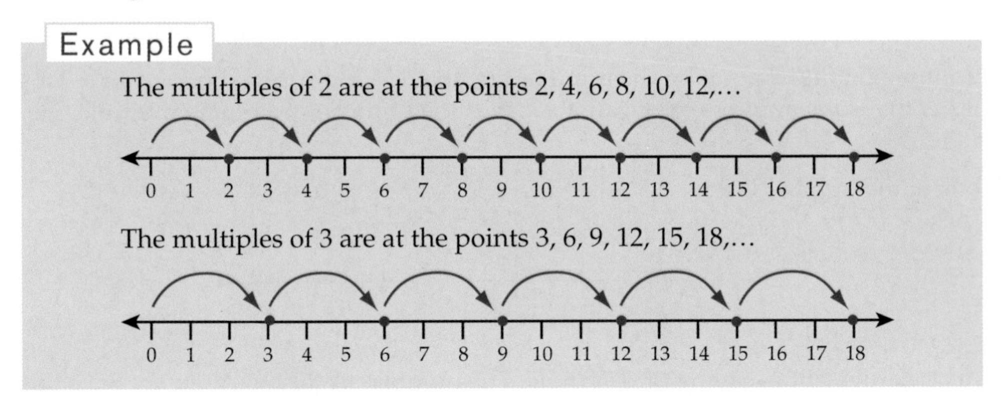

Common Multiples

If two numbers have the same multiple, you call the multiple a *common multiple* of those two numbers.

Here are some ways to identify common multiples.

Common Multiples in the Multiplication Table

Look at the "6" column of the small multiplication table on page 132. You can see that 12 is a multiple of 6. Now, look at the "3" column. You can see that 12 is also a multiple of 3. So 12 is a common multiple of 6 and 3.

Imagine a giant multiplication table that goes on forever. If you looked in the "6" column, you would see 90 in the "15" row. You would also see 90 in the "15" column of row "6."

When a number appears in two different columns, it is a *common multiple* of the numbers at the top of each column.

Common Multiples on the Number Line

If you locate the multiples of a on the number line, they will be at a, $2a$, $3a$, $4a$,..., ma,..., and so on.

Likewise, the multiples of b are located at b, $2b$, $3b$, $4b$,..., nb,..., and so on.

Suppose you find a point on the number line that is a multiple of both a and b. This means that some multiple of a, call it ma, is at the same point as some multiple of b, call it nb.

Look at the number line:

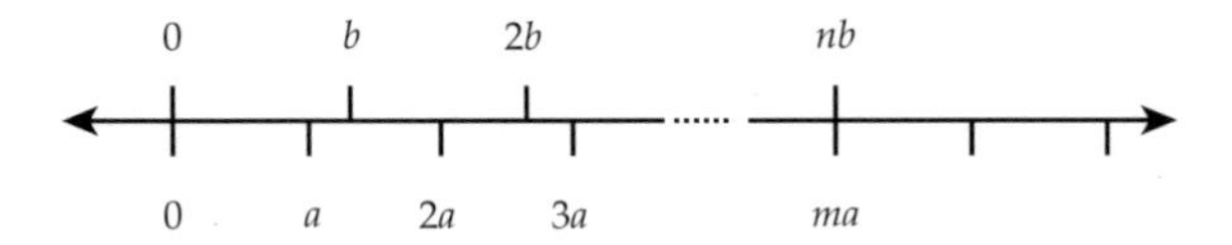

This is equivalent to saying that $ma = nb$. The number ma (also written nb, since they are equal in this case) is a common multiple of a and b.

Example

To find the common multiples of 2 and 3, you can compare the two number lines of their multiples or compare the lists of multiples.

The multiples of 2 are 2, 4, 6, 8, 10, 12, 14, 16, 18,...

The multiples of 3 are 3, 6, 9, 12, 15, 18,...

The common multiples of 2 and 3 are 6, 12, 18,...

The common multiples of 2 and 3 are the multiples of 6.

Common Multiples Using Letters

If $a \bullet b = c$, where a, b, and c are natural numbers, then c is a multiple of a, and c is a multiple of b. Therefore, all the multiples of c are common multiples of both a and b.

Example

Let $a = 8$, and $b = 10$
$a \bullet b = c$
$8 \bullet 10 = c$
$80 = c$
80 is a common multiple of both 8 and 10.

Some of the common multiples of 8 and 10 are $80n$:
80, 160, 240, 320, 400, 480,…

Least Common Multiple

You have looked at common multiples of a and b that are equal to or greater than the product $a \bullet b$. Sometimes, there is a common multiple that is less than the product of the factors. In other words, there are common multiples of a and b that are less than $a \bullet b$.

Example

40 is a common multiple of both 8 and 10 and is less than 80, the product of the factors 8 and 10.

$40 = 8 \bullet 5$ and $40 = 10 \bullet 4$

The expression $40 \bullet n$ represents additional common multiples of 8 and 10:
40, 80, 120, 160, 200, 240, 280,…

The common multiple of two numbers with the smallest value is called the *least common multiple*. The least common multiple is either less than or equal to the product of the two numbers. The least common multiple is sometimes shortened to LCM.

The least common multiple can be found by generating lists of multiples and then finding the smallest number among them.

Example

To find the least common multiple of 8 and 10, list the multiples of 8 and 10:

Multiples of 8: 8, 16, 24, 32, (40), 48, 56, 64, 72, 80,...

Multiples of 10: 10, 20, 30, (40), 50, 60, 70, 80,...

The smallest number in both lists is 40, which is the least common multiple of 8 and 10.

Using Prime Factorization to Find the Least Common Multiple

Another way to find the least common multiple is by factoring into primes.

Example

Find the least common multiple of 6 and 15. Look at the prime factorizations of 6 and 15 and their product, 90.

$6 = 2 \cdot 3$ $\quad\quad 15 = 3 \cdot 5$ $\quad\quad 90 = 2 \cdot 3 \cdot 3 \cdot 5$

Any number that can be written as $2 \cdot 3 \cdot n$, for some number n, is a multiple of 6. Similarly, any number that can be written as $3 \cdot 5 \cdot m$, for some number m, is a multiple of 15.

Now, notice that the number $2 \cdot 3 \cdot 5$ fits into both the equations above, since it has both a factor $2 \cdot 3$ and also a factor $3 \cdot 5$. Therefore, $2 \cdot 3 \cdot 5 = 30$ is a common multiple of both 6 and 15.

If you check this, you get: $30 = 6 \cdot 5$ and $30 = 15 \cdot 2$.

So, 30 is a smaller common multiple of 6 and 15 than 90.
In fact, 30 is the least common multiple of 6 and 15.

There is a very important idea in what you have just done. In looking for a common multiple of numbers like $6 = 2 \bullet 3$ and $15 = 3 \bullet 5$ that have a *common factor* (in this case, the factor 3), you only need to use that common factor once in creating a common multiple.

In general, to find the least common multiple of two, three, or more numbers:

1. Start with the prime factorization of the first number.
2. Multiply this by any prime factors in the second number that are not already included in the prime factorization of the first number.
3. Multiply this by any prime factors in the third number that are not already included in the prime factorization of the second number.
4. Continue this process for all remaining prime factors for each number.

Example

What is the least common multiple of the numbers 15, 20, 36, and 210?

1. Here are the prime factorizations of the numbers:

   $15 = 3 \bullet 5$ $\quad$ $20 = 2 \bullet 2 \bullet 5$

   $36 = 2 \bullet 2 \bullet 3 \bullet 3$ $\quad$ $210 = 2 \bullet 3 \bullet 5 \bullet 7$

   Start with the prime factorization of 15: $3 \bullet 5$.

2. Multiply this by $2 \bullet 2$ (the part of the prime factorization of 20 not already included).

   This gives $3 \bullet 5 \bullet 2 \bullet 2$.

3. Multiply this by 3 (the part of the prime factorization of 36 not already included in the prime factorization of 15 or 20).

   This gives $3 \bullet 5 \bullet 2 \bullet 2 \bullet 3$.

4. Multiply this by 7 (the part of the prime factorization of 210 not already included).

   This gives $3 \bullet 5 \bullet 2 \bullet 2 \bullet 3 \bullet 7$. You are done.

Example

The number $2 \bullet 2 \bullet 3 \bullet 3 \bullet 5 \bullet 7 = 1{,}260$ has the factors necessary to be a multiple of each of 15, 20, 36, and 210:

$$3 \bullet 5 \text{ times } 2 \bullet 2 \bullet 3 \bullet 7 = 15 \bullet 84 = 1{,}260$$

$$2 \bullet 2 \bullet 5 \text{ times } 3 \bullet 3 \bullet 7 = 20 \bullet 63 = 1{,}260$$

$$2 \bullet 2 \bullet 3 \bullet 3 \text{ times } 5 \bullet 7 = 36 \bullet 35 = 1{,}260$$

$$2 \bullet 3 \bullet 5 \bullet 7 \text{ times } 2 \bullet 3 = 210 \bullet 6 = 1{,}260$$

The number 1,260 is the least common multiple of 15, 20, 36 and 210, because if you leave out any one of the factors, the result would not be a multiple of each of these numbers.

That is, you cannot do without any one of the factors to make a smaller common multiple than 1,260.

$$\text{LCM } \{15, 20, 36, 210\} = 2 \bullet 2 \bullet 3 \bullet 3 \bullet 5 \bullet 7 = 1{,}260$$

In general, the following equation is true:

$$(\text{GCF of } a \text{ and } b) \bullet (\text{LCM of } a \text{ and } b) = a \bullet b$$

Example

If $a = 6$, $b = 15$

The GCF of 6 and 15 is 3.

The LCM of 6 and 15 is 30.

$\text{GCF} \bullet \text{LCM} = a \bullet b$

$3 \bullet 30 = 6 \bullet 15$

$90 = 90$

FRACTIONS AND DECIMALS

CHAPTER 9

Representing Fractions

A *fraction* uses a fraction bar to represent a number as a division of two natural numbers, m and n. $\frac{m}{n}$ is a fraction; n cannot be 0. m is called the numerator; n is called the denominator.

Example

$\frac{3}{4}$ is a fraction. 3 is the numerator; 4 is the denominator.

$\frac{22}{7}$ is fraction. 22 is the numerator; 7 is the denominator.

The *fraction bar* indicates division: $\frac{3}{4}$ or $3 \div 4$.

To find the decimal form of a fraction, you simply carry out the division, either by hand or with a calculator.

The numerator is divided by the denominator:

$$\frac{\text{numerator}}{\text{denominator}} \text{ or } \text{denominator}\overline{\smash{)}\text{numerator}}$$

Example

The fraction $\frac{7}{8}$ means 7 divided by 8.

The fraction $\frac{7}{8}$ is equal to the decimal number 0.875.

$$
\begin{array}{r}
0.875 \\
8\overline{)7.000} \\
\underline{-64} \\
60 \\
\underline{-56} \\
40 \\
\underline{-40} \\
0
\end{array}
$$

The fraction $\frac{3}{4}$ is equal to 0.75.

The fraction $\frac{22}{7}$ is equal to 3.142857142857…

Generally, a fraction is a representation of the form $\frac{m}{n}$, where m and n are natural numbers. This includes:

- Proper fractions, where $m < n$
- Improper fractions, where $m > n$
- Whole numbers, where $n = 1$ or $m = n$

Example

The fraction $\frac{7}{8}$ is a proper fraction because $7 < 8$.

The fraction $\frac{22}{7}$ is an improper fraction because $22 > 7$.

$\frac{2}{1} = 2$ is a whole number.

Here is part of a number line that represents some of the numbers between 0 and 1 as decimals and as fractions.

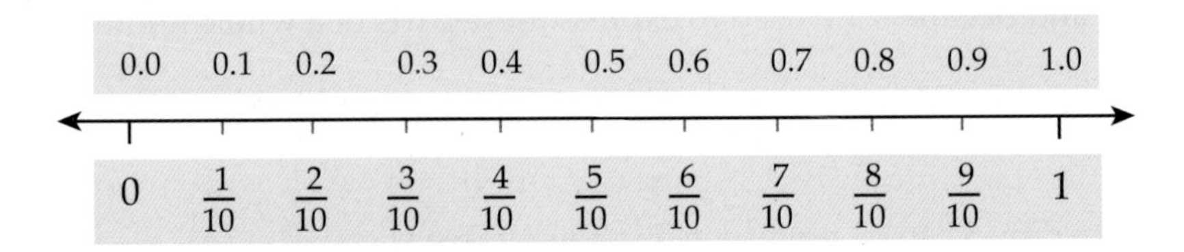

A number can also be expressed as a mixed number. A *mixed number* is a number consisting of the sum of an integer and a fraction.

Example

The number $3\frac{1}{2}$ is a mixed number. 3 is the integer part and $\frac{1}{2}$ is the fraction part. It can be written as a decimal, 3.5, or as an *improper fraction*, $\frac{7}{2}$, or as a sum, $3 + \frac{1}{2}$.

Here is another part of the number line. This part represents some of the numbers between 0 and 6 as mixed numbers, fractions, and decimals.

0	$\frac{1}{2}$	1	$1\frac{1}{2}$	2	$2\frac{1}{2}$	3	$3\frac{1}{2}$	4	$4\frac{1}{2}$	5	$5\frac{1}{2}$	6
$\frac{0}{2}$	$\frac{1}{2}$	$\frac{2}{2}$	$\frac{3}{2}$	$\frac{4}{2}$	$\frac{5}{2}$	$\frac{6}{2}$	$\frac{7}{2}$	$\frac{8}{2}$	$\frac{9}{2}$	$\frac{10}{2}$	$\frac{11}{2}$	$\frac{12}{2}$
0.0	0.5	1.0	1.5	2.0	2.5	3.0	3.5	4.0	4.5	5.0	5.5	6.0

Reading and Writing Fractions

Measurements can be made to an accuracy that is greater than the nearest whole number. Fractions and decimals are used to express these parts of a whole number

Fractions and decimals can also represent one natural number divided by another.

Example

12 divided by 5 can be represented in these three ways:

as $\frac{\text{numerator}}{\text{denominator}}$: $\frac{12}{5}$	You say, "twelve-fifths" or "twelve over five."
as a mixed number: $2\frac{2}{5}$ (part integer and part fraction)	You say, "two and two-fifths."
as a decimal: 2.4	You say, "two and four-tenths" or "two point four."

$\frac{12}{5}$, $2\frac{2}{5}$, and 2.4 are all equal.

They are *equivalent* representations of the same number.

They all indicate the same position on the number line.

Fractions on the Number Line

On the number line, integers are shown as points spaced at equal distances, or intervals.

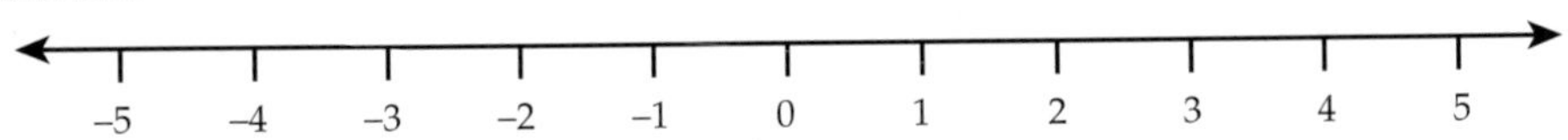

To show fractions between 0 and 1 with denominator 5, the space between the whole number points 0 and 1 must be divided into five equal parts, or fifths.

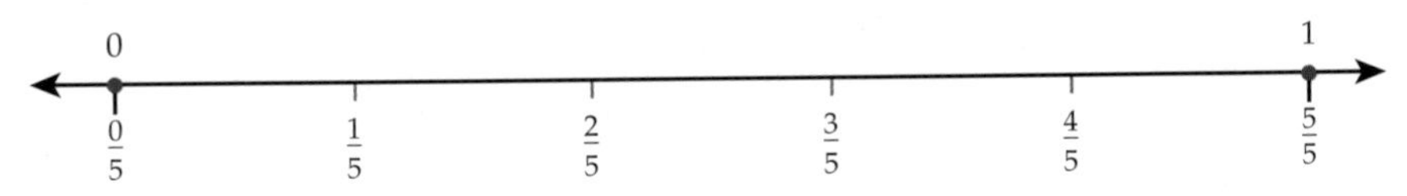

To show fractions between 0 and 1 with denominator 10, the space between the whole number points 0 and 1 must be divided into ten equal parts, or tenths.

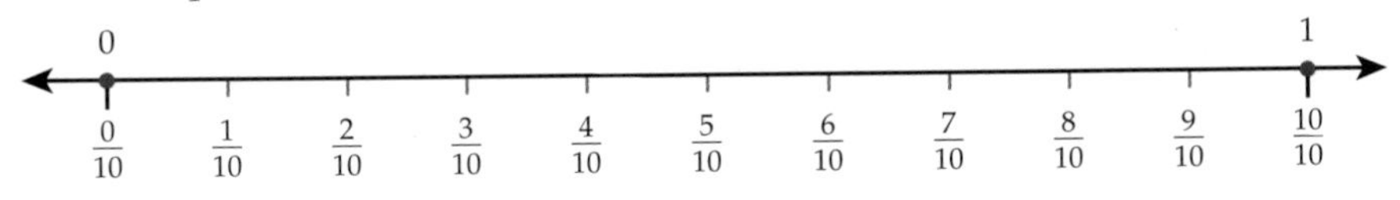

There is an infinite number of numbers between any two points on the number line. This is true because you can always find a point halfway between two points.

Example

The number halfway between 2.4 and 2.5 is 2.45, or $2\frac{9}{20}$.

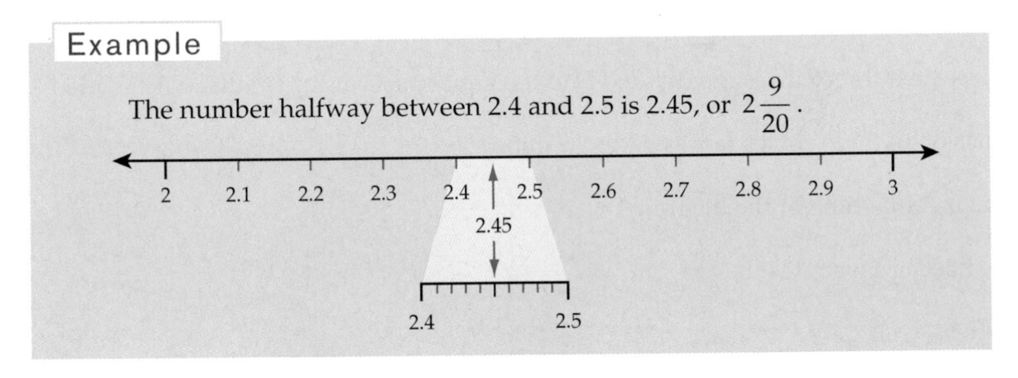

Chapter 9

Putting Fractions in Order

Every number has a definite position on the number line.

Greater than means "further to the right" on the horizontal number line. The symbol for "is greater than" is > and "is greater than or equal to" is ≥.

Less than means "further to the left" on the horizontal number line. The symbol for "is less than" is < and "is less than or equal to" is ≤.

Example

$2\frac{9}{20}$ is greater than $2\frac{2}{5}$. You write: $2\frac{9}{20} > 2\frac{2}{5}$.

$2\frac{9}{20}$ is less than $3\frac{1}{2}$. You write: $2\frac{9}{20} < 3\frac{1}{2}$.

Representing Fractions with Circles

Circles divided into parts can illustrate fractions. A "whole" circle represents the number 1, and parts of the circle represent fractions less than 1.

Here, whole circles are divided into five equal parts, or into fifths. The shaded parts of the three circles represent the number $2\frac{2}{5} = \frac{12}{5}$.

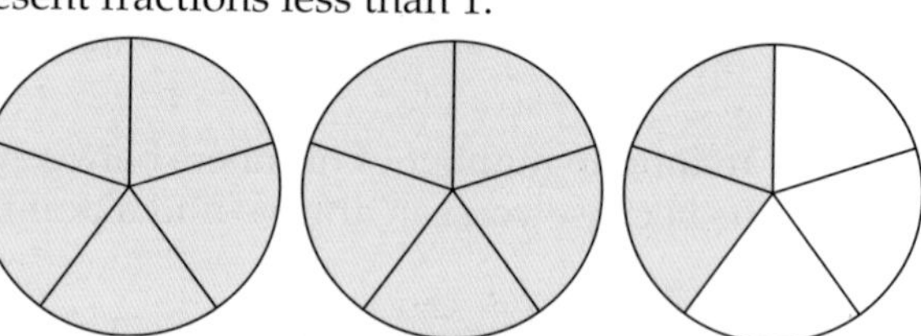

These next three circles are divided into ten equal parts or into tenths. The shaded parts of all three circles represent the number $2\frac{4}{10} = \frac{24}{10}$.

The fact that both of the diagrams have the same amount of shading illustrates the fact that $2\frac{2}{5} = 2\frac{4}{10} = 2.4$.

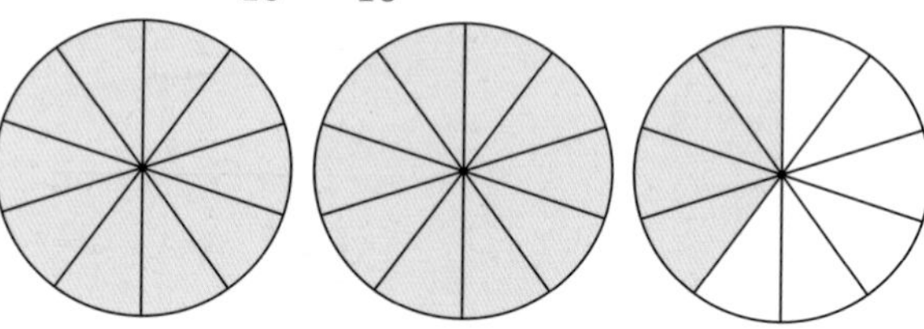

Equivalent Fractions

The fractions $\frac{12}{5}$ and $\frac{24}{10}$ are examples of *equivalent fractions*.

Equivalent fractions are equal. You write $\frac{12}{5} = \frac{24}{10}$.

Equivalent fractions have the same position on the number line.
They are the same distance from 0.

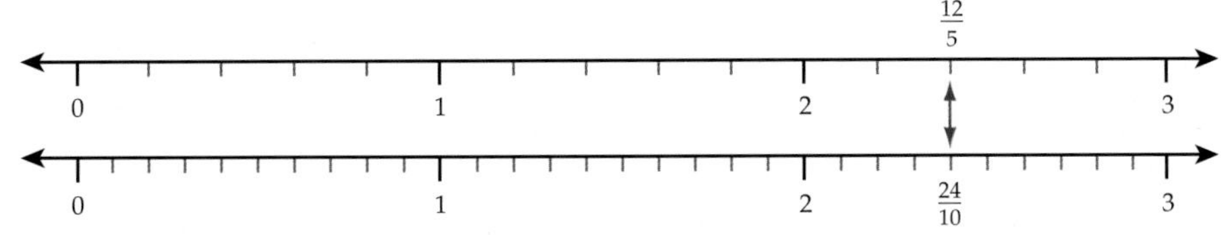

On the first number line, each unit is broken into 5 equal parts (fifths).

In $\frac{12}{5}$, the numerator, 12, represents the *number* of parts and the denominator, 5, indicates the *size* of the parts (fifths).

On the second number line, each unit is broken into 10 equal parts (tenths).

In $\frac{24}{10}$, the numerator, 24, represents the *number* of parts and the denominator, 10, indicates the *size* of the parts (tenths).

A large denominator means that each whole unit has been divided into small parts. A small denominator means that each whole unit has been divided into large parts.

Example

Think about money. A quarter is $\frac{1}{4}$ of a dollar. A dime is $\frac{1}{10}$ of a dollar. The quarter is worth more than the dime—it is a bigger part of a dollar. The quarter has a smaller denominator (4) than the dime (10).

If f is any natural number, then $\frac{a \bullet f}{b \bullet f}$ and $\frac{a}{b}$ are equal: $\frac{a}{b} = \frac{a \bullet f}{b \bullet f}$. They are called *equivalent fractions.*

This is true because the value of the fraction $\frac{f}{f}$ is 1, which is the multiplicative identity.

Look back at $\frac{12}{5}$. If you let $f = 2$, then $\frac{12}{5} = \frac{12 \bullet 2}{5 \bullet 2} = \frac{24}{10}$. So $\frac{12}{5}$ is equivalent to $\frac{24}{10}$.

The first fraction $\frac{12}{5}$ is called a *reduced form* of $\frac{24}{10}$. Since you cannot reduce $\frac{12}{5}$ any further, you say that $\frac{12}{5}$ is the *simplest form* of $\frac{24}{10}$.

The simplest form of a fraction is when the numerator and denominator have no factors in common other than 1.

Example

$0.75 = \frac{75}{100}$

Since 5 is a factor of both 75 and 100, you can see that:

$$0.75 = \frac{75}{100} = \frac{15 \bullet 5}{20 \bullet 5} = \frac{15}{20}$$

$\frac{15}{20}$ is a reduced form, but it is not the simplest form.

5 is a factor of 15 and 20, so $\frac{15}{20} = \frac{3 \bullet 5}{4 \bullet 5} = \frac{3}{4}$.

So, $\frac{3}{4}$ is a reduced form of $\frac{75}{100}$.

In fact, $\frac{3}{4}$ is the simplest form of $\frac{75}{100}$.

Adding and Subtracting Fractions

Fractions with a Common Denominator

The fractions $\frac{8}{5}$ and $\frac{12}{5}$ are shown on the following number line.

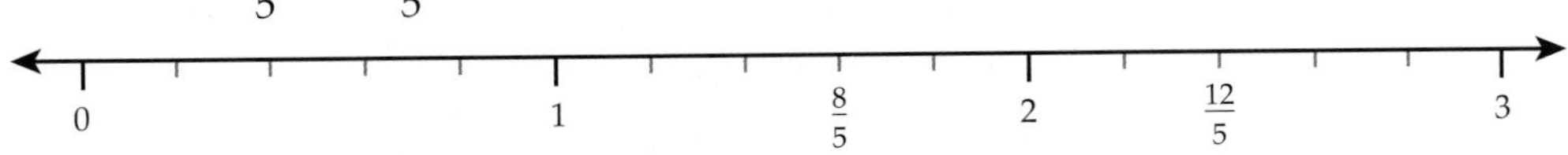

The denominators of 5 indicate that each interval is equal to a size of $\frac{1}{5}$.

The numerators of 8 and 12 indicate the number of steps, starting from 0. These same-size steps can be added and subtracted.

$\frac{8}{5}$ is four steps to the left of $\frac{12}{5}$, so the distance between these two numbers is $\frac{12}{5} - \frac{8}{5} = \frac{4}{5}$.

Similarly, $\frac{12}{5}$ is four steps forward from $\frac{8}{5}$, thus $\frac{8}{5} + \frac{4}{5} = \frac{12}{5}$.

Summary

To add or subtract fractions with a common denominator, just add or subtract the numerators, and leave the denominators unchanged.

In symbols, $\frac{a}{c} + \frac{b}{c} = \frac{a+b}{c}$ and $\frac{a}{c} - \frac{b}{c} = \frac{a-b}{c}$.

Fractions with Different Denominators

To add or subtract fractions with different denominators, you can create equivalent fractions that have the same denominators.

Example

To use the number line to find the distance from $\frac{13}{10}$ to $\frac{12}{5}$, you must first put both fractions on a number line with the same scale; in this case, the scale marked in tenths. Note that $\frac{12}{5} = \frac{24}{10}$.

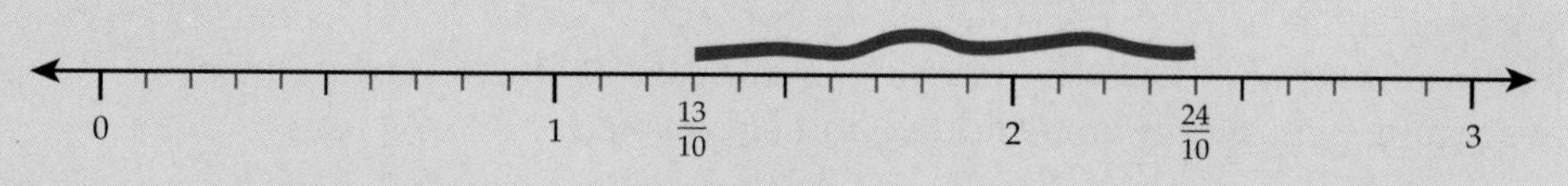

Thus, the distance is $\frac{12}{5} - \frac{13}{10} = \frac{24}{10} - \frac{13}{10} = \frac{24 - 13}{10} = \frac{11}{10}$.

Example

To add $\frac{12}{5}$ and $\frac{3}{7}$, a common denominator could be $5 \bullet 7 = 35$.

So, $\frac{12}{5} = \frac{12 \bullet 7}{5 \bullet 7} = \frac{84}{35}$ and $\frac{3}{7} = \frac{3 \bullet 5}{7 \bullet 5} = \frac{15}{35}$.

Adding then gives $\frac{12}{5} + \frac{3}{7} = \frac{84}{35} + \frac{15}{35} = \frac{99}{35}$.

Summary

In symbols: $\frac{a}{b} + \frac{c}{d} = \frac{a \bullet d}{b \bullet d} + \frac{c \bullet b}{d \bullet b} = \frac{(a \bullet d) + (c \bullet b)}{b \bullet d}$

Multiplying a Fraction by a Whole Number

Multiplying a fraction by a whole number can be represented as repeated addition on the number line.

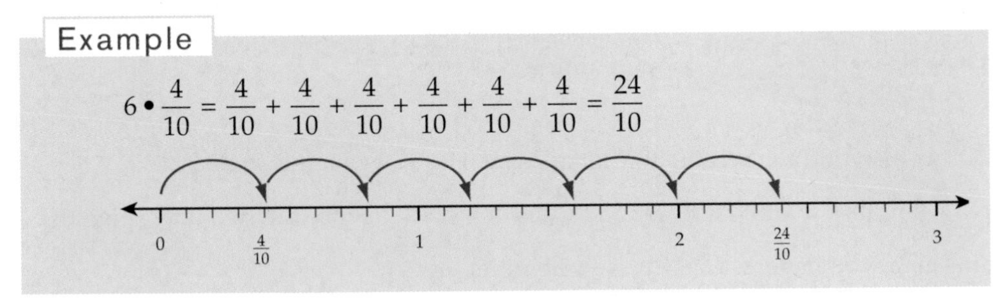

In symbols: $n \bullet \frac{a}{b} = \frac{n \bullet a}{b}$

Multiplying Two Fractions

The diagram shows two number lines at right angles to each other. The vertical line is marked in halves. The horizontal line is marked in quarters.

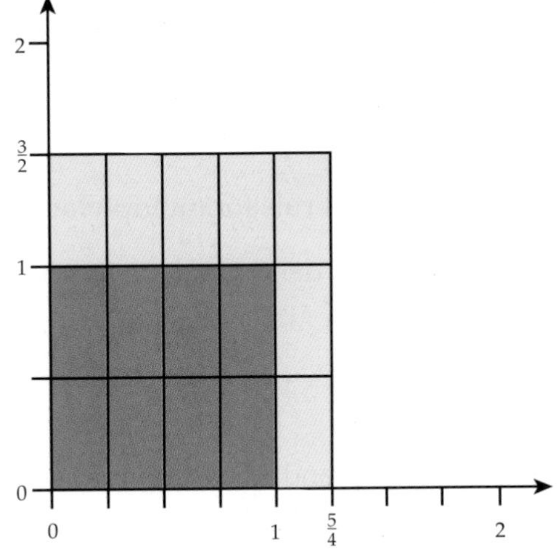

The "unit square" is shown with darker shading. Its sides go from 0 to 1 on each number line.

The "unit square" has an area of $1 \bullet 1 = 1$ square unit. You see that the unit square has $2 \bullet 4 = 8$ small rectangles. So the unit square has been divided into 8 smaller rectangles. Since these rectangles are all the same, each has an area of $1 \div 8 = \frac{1}{8}$ square units.

What is the area of the largest shaded rectangle?

The largest rectangle contains 3 rows and 5 columns of rectangles, each with an area of $\frac{1}{8}$ square unit. Thus, the whole area of the largest shaded rectangle is $(3 \bullet 5) \bullet \frac{1}{8} = 15 \bullet \frac{1}{8} = \frac{15}{8}$ square units.

You can also find the area in a different way. The area of a rectangle is area = length • width. Here, length • width = $\frac{5}{4} \bullet \frac{3}{2}$. Since the two ways give the same area, you have $\frac{5}{4} \bullet \frac{3}{2} = \frac{15}{8}$ square units.

> The numerator in the product $\frac{15}{8}$ is the total number of small rectangles, $3 \bullet 5 = 15$.
>
> The denominator in the product $\frac{15}{8}$ is the number of small rectangles in the unit square, $2 \bullet 4 = 8$.

The general formula shows that the product of any two fractions, $\frac{a}{b}$ and $\frac{c}{d}$, is $\frac{a \bullet c}{b \bullet d}$.

Summary

Thus, the general rule for multiplying fractions is simply to multiply the numerators and multiply the denominators. In symbols: $\frac{a}{b} \bullet \frac{c}{d} = \frac{a \bullet c}{b \bullet d}$

Reciprocals

The numbers $\frac{a}{b}$ and $\frac{b}{a}$, with numerators and denominators inverted, are called *reciprocals* of each other.

The key property of reciprocals is that their product is 1, since $\frac{a}{b} \bullet \frac{b}{a} = \frac{a \bullet b}{a \bullet b} = 1$.

Reciprocals are called *multiplicative inverses* of each other because the effect of multiplying by a fraction $\frac{a}{b}$ is undone by then multiplying the result by the reciprocal, $\frac{b}{a}$.

If a fraction is between 0 and 1, its reciprocal will be greater than 1. If a number is greater than 1, its reciprocal will be less than 1.

Example

Fraction	Reciprocal
$\frac{1}{2} < 1$	$\frac{2}{1} > 1$
$\frac{3}{10} < 1$	$\frac{10}{3} > 1$
$\frac{8}{100} < 1$	$\frac{100}{8} > 1$
$2 > 1$	$\frac{1}{2} < 1$
$3\frac{1}{3} > 1$	$\frac{3}{10} < 1$

In symbols, for any number f: $\left(f \bullet \frac{a}{b}\right) \bullet \frac{b}{a} = f \bullet \left(\frac{a}{b} \bullet \frac{b}{a}\right) = f \bullet 1 = f$

Dividing Fractions

A multiplication that represents *area of rectangle = length times width (A = lw)* is equivalent to a division that represents:

area of rectangle divided by width = length $\left(\frac{A}{w} = l\right)$

Example

Suppose you know that the area of a rectangle is $\frac{15}{8}$ square units, and that the width is $\frac{3}{2}$ units. Then:

$$\text{length} = \frac{\text{area}}{\text{width}} = \frac{15}{8} \div \frac{3}{2}$$

To find the length you have to divide one fraction, $\frac{15}{8}$, by another fraction, $\frac{3}{2}$.
Use an area model to help you divide.

The length of the side you want to find is marked with a question mark.

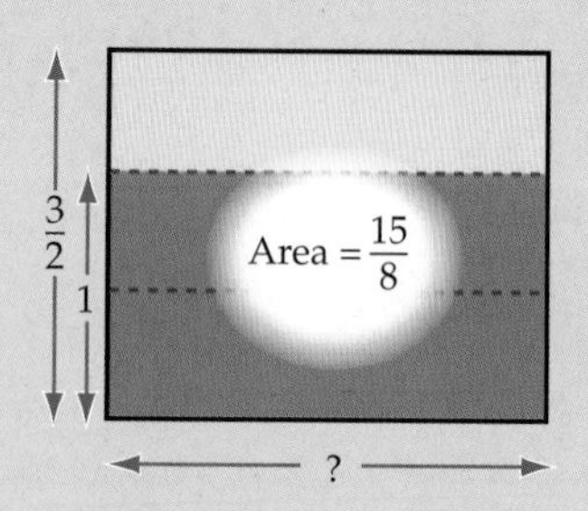

The lower two rectangles have $\frac{2}{3}$ of the area and a width of 1. Since the length is the same,

$$\text{length} = \frac{\text{area}}{\text{width}} = \left[\frac{15}{8} \bullet \frac{2}{3}\right] \div 1 = \frac{15}{8} \bullet \frac{2}{3}$$

(because 1 is the identity element)

Dividing by $\frac{3}{2}$ is equivalent to multiplying by $\frac{2}{3}$:

$$\text{length} = \frac{15}{8} \div \frac{3}{2} = \frac{15}{8} \bullet \frac{2}{3} = \frac{30}{24} = \frac{5}{4}$$

Example

$4 \div \frac{2}{5}$

How many times does $\frac{2}{5}$ go into 4?

You can illustrate this problem by cutting 4 squares into fifths.

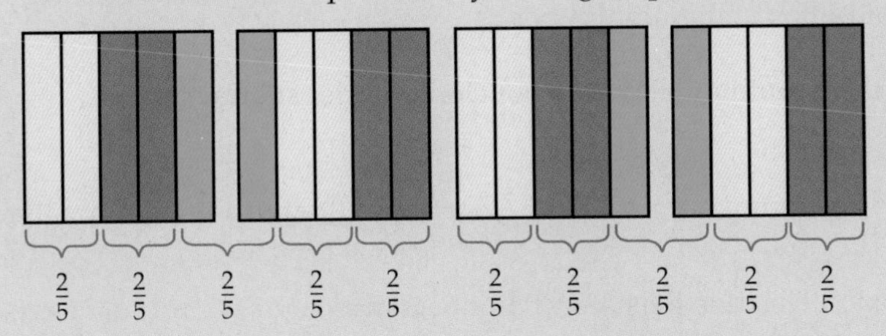

Four squares are made of ten parts, with each part being $\frac{2}{5}$. Thus, $4 \div \frac{2}{5} = 10$.

If you take to be one unit, how many units make up one square?

Each square is made of $2\frac{1}{2} = \frac{5}{2}$ of these units. Since there is a total of 4 squares, the total number of units is $4 \bullet \frac{5}{2} = \frac{20}{2} = 10$.

Thus, dividing by $\frac{2}{5}$ is equivalent to multiplying by $\frac{5}{2}$. Therefore:

$$4 \div \frac{2}{5} = 4 \bullet \frac{5}{2} = \frac{20}{2} = 10.$$

Summary

The general rule is: *To divide by a fraction, multiply by the reciprocal of the fraction.*

In symbols: $\frac{a}{b} \div \frac{c}{d} = \frac{a}{b} \bullet \frac{d}{c} = \frac{a \bullet d}{b \bullet c}$

Types of Numbers

The *natural numbers* are the numbers you use for counting. They start at 1 and continue forever.

When you add or multiply natural numbers together, you always get another natural number as the result. You describe this property by saying, "The natural numbers are *closed* under addition and multiplication."

By contrast, the natural numbers are not closed under subtraction.

Example

5 is a natural number. But $5 - 5 = 0$, and 0 is not a natural number.

The *whole numbers* include the natural numbers plus zero. The role of zero is most important in our place-value system of writing numbers. Zero shows that there is nothing to count in that place.

Example

The 0 in the number 3,045 means that there are no hundreds to be counted.

When you add or multiply two whole numbers you always get another whole number. You say, "The whole numbers are closed under addition and under multiplication."

The *integers* include:

- The natural numbers {1, 2, 3,…}. These are called the *positive integers*
- Their opposites, the *negative integers* {–1, –2, –3,…}
- *Zero*, which is neither positive nor negative

You can add, subtract, or multiply any two integers and the result will always be another integer. You say, "The integers are closed under addition, subtraction, and multiplication."

The *rational numbers* are the numbers that result when one integer is divided by another integer (not equal to 0).

Example

$\frac{5}{2}, \frac{2}{1}, \frac{-7}{10}, \frac{0}{-3}$ are rational numbers.

The rational numbers are closed for all the basic operations of arithmetic—addition, subtraction, multiplication, and division—with the exception that you cannot divide by 0.

Some numbers, such as π and $\sqrt{2}$, cannot be represented in the form $\frac{m}{n}$, where m and n are integers. These numbers are not rational numbers. They are called *irrational numbers,* and together with the rational numbers, they make up the whole set of numbers called *real numbers.*

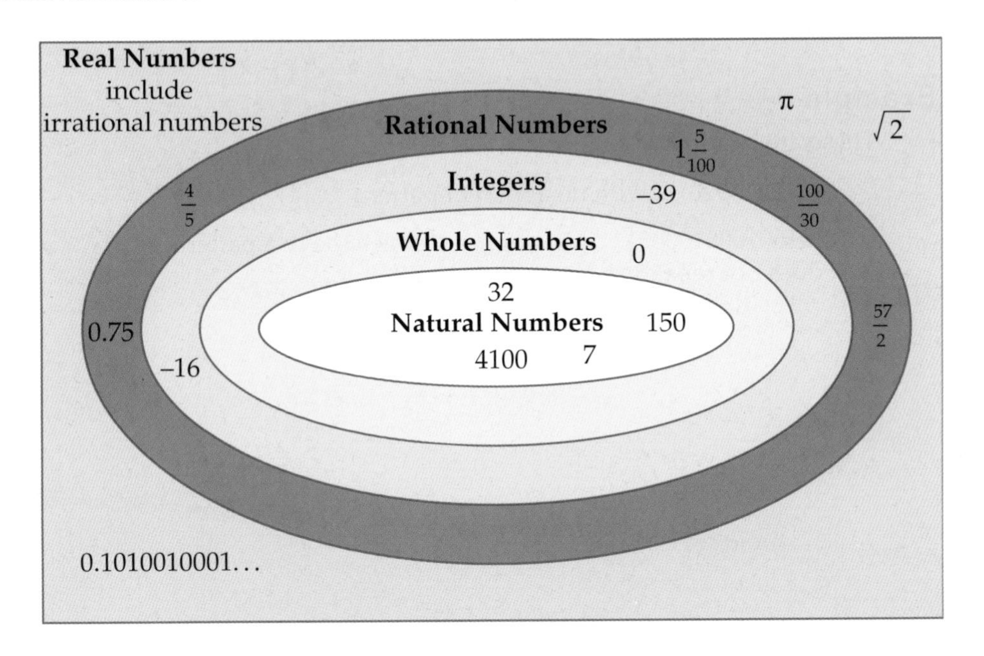

More About Rational Numbers

A rational number is any number that can be written in the form $\frac{a}{b}$, where a and b are integers, and where $b \neq 0$.

The expression $\frac{0}{2}$ is one of the many ways of writing zero.

The expression $\frac{5}{0}$ has no meaning as a number, since it is impossible to divide by zero.

As decimals, all rational numbers are either:

- Terminating decimals, such as $\frac{7}{4} = 1.75$
- Repeating decimals, such as $\frac{3}{11} = 0.272727\ldots = 0.\overline{27}$
  The bar over the 27 means it repeats forever.

Irrational numbers are neither terminating nor repeating decimals.

Example

These are irrational numbers:

$\pi = 3.1415926\ldots$, with no apparent pattern

0.101001000100001…, although it seems to have a pattern, it does not repeating.

More About Irrational Numbers

Decimals that do not terminate or repeat are called *irrational numbers. Irrational* means "not rational."

Example

- The number π is an irrational number. A decimal approximation for π is 3.1415926… No matter how many decimal places you go to, there is no repeating pattern. You cannot write an exact decimal value for π.
- Another irrational number is 0.123456789101112… The pattern in these digits could go on forever, but it is not a repeating pattern, so the number is not a rational number.
- Another irrational number is $\sqrt{2}$. While 1.41 is an approximation of $\sqrt{2}$, there is no rational number that when multiplied by itself has a product of exactly 2.

There is an infinite number of irrational numbers. There is also an infinite number of rational numbers. The rational numbers and the irrational numbers make up the *real numbers*. Every real number has a unique position on the number line.

Using Rational Numbers to Measure

All measurements are approximations. When a rational number is used for the value of a measurement, the denominator tells you how precise the measurement is.

One end of this piece of wire has been placed as close as possible to 0.

If you read the scale at the other end to the nearest inch, the measurement is 2 inches.

If you read the scale to the nearest $\frac{1}{2}$ inch, the measurement is still $\frac{4}{2} = 2$ inches, since the reading is closer to 2 than to $2\frac{1}{2}$.

If you read the scale to the nearest $\frac{1}{4}$ inch, the measurement is $2\frac{1}{4}$ inches, since the reading is closer to $2\frac{1}{4}$ than to 2.

If you read the scale to the nearest $\frac{1}{8}$ inch, the measurement is $2\frac{1}{8}$ inches, since the reading is closer to $2\frac{1}{8}$ than to $2\frac{2}{8}$.

If you read the scale to the nearest $\frac{1}{16}$ inch, the measurement is $2\frac{2}{16}$ inches, since the reading is closer to $2\frac{2}{16}$ than to $2\frac{3}{16}$.

Estimation

To *estimate* a value means to make an informed guess, or to give an approximate measurement.

Estimating Square Roots

Trial and improvement is a process of making informed guesses and using the results of each guess to make a better guess.

Example

If the area of a square is 11 m^2, what is the length of a side?

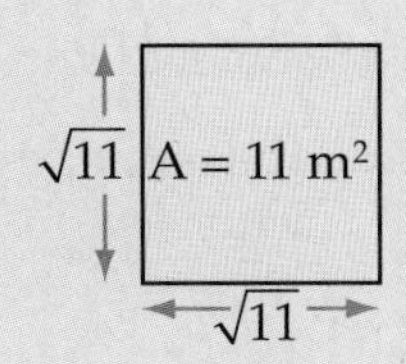

The area of a square equals the length of a side squared.

You must ask yourself, "What number squared equals 11?" Each side must be the square root of 11.

To find $\sqrt{11}$, try numbers until you get close to a product of 11.

$3 \cdot 3 = 9$	too small
$4 \cdot 4 = 16$	too large
$3.5 \cdot 3.5 = 12.25$	too large
$3.4 \cdot 3.4 = 11.56$	too large
$3.3 \cdot 3.3 = 10.89$	too small

The answer must lie between 3.3 and 3.4, and appears to be closer to 3.3.

$3.31 \cdot 3.31 = 10.9561$ too small
3.31 is a close estimate of $\sqrt{11}$.

You could continue to use any number of decimal places to estimate a more precise value of any square root.

Making Measurements

All measurements are approximations. When a decimal number represents a measurement, the number of decimal places tells you how precise that measurement is.

Example

An object that is measured as 5.4 meters long is actually somewhere between 5.35 and 5.45 meters long.

If the same object is measured as 5.4001 m, the range of possible actual lengths would be much narrower. The actual length would be somewhere between 5.40005 m and 5.40015 m.

The second measurement, 5.4001 m, is a more precise estimate of the actual length.

Rounding

Calculators sometimes round numbers when they cannot show all of the digits of a number. The calculator rounds this way with repeating decimals and with nonterminating, nonrepeating decimals. Calculators also round very long terminating decimals.

Example

The repeating decimal 0.6666666666… might be rounded by a calculator to 0.66666667 or to 0.666666666666667, depending on the number of digits shown on the display.

Example

The number 852.6543219876543215897654321 might be rounded to 852.6543219877.

In the first example, you can see that the number is most likely a repeating decimal. In the second, you cannot be sure whether the number terminates after the second 7, continues on and terminates at another digit, or repeats.

Converting a Fraction to a Decimal

To convert a fraction to decimal form, divide the numerator by the denominator.

$$\frac{\text{numerator}}{\text{denominator}} \text{ or } \text{denominator}\overline{)\text{numerator}}$$

Example

To convert $\frac{5}{8}$ to a decimal, write the 5 as 5.000 and do this division:

$$\begin{array}{r} 0.625 \\ 8\overline{)5.{}^{5}0{}^{2}0{}^{4}0} \\ -4.8 \\ \hline 20 \\ -16 \\ \hline 40 \\ -40 \\ \hline 0 \end{array}$$

Terminating and Repeating Decimals

When you divide one integer by a nonzero integer, one of two interesting things can happen.

The number might be represented by a *terminating* decimal, or the number might be represented by a *repeating* decimal.

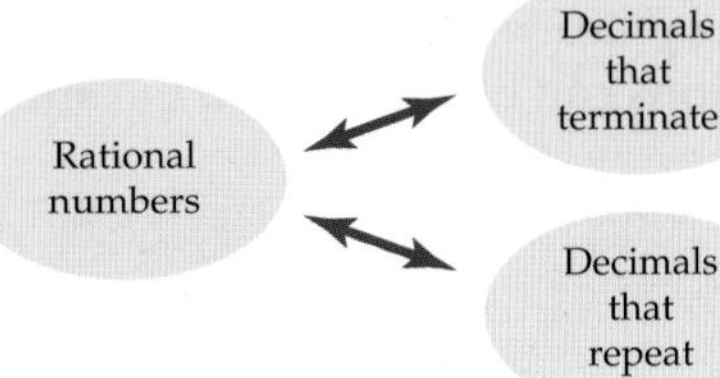

A *terminating decimal* stops after a finite number of digits. Any further digits are zeros.

Example

$\frac{3}{8}$ is a terminating decimal.

The decimal 0.375 stops after three digits because the third place divides exactly with no remainder.

$$\begin{array}{r} 0.375 \\ 8\overline{)3.000} \\ \underline{-2.4} \\ 60 \\ \underline{-56} \\ 40 \\ \underline{-40} \\ 0 \end{array}$$

Example

These numbers are examples of terminating decimals.

$\frac{4}{5} = 4 \div 5 = 0.8$

$-\frac{375}{1000} = -375 \div 1{,}000 = -0.375$

Using place value, terminating decimals can be expressed as fractions. Then the fractions can be reduced to the simplest form of the fraction.

Example

$0.656 = \frac{656}{1000} = \frac{82 \bullet 8}{125 \bullet 8} = \frac{82}{125}$

$2.14 = 2\frac{14}{100} = 2\frac{7}{50}$

When the quotient $a \div b$ is represented as a decimal with a finite number of digits that repeat infinitely, it is called a *repeating decimal*.

Example

$\frac{5}{11} = 0.454545\ldots$ is a repeating decimal.

The decimal never terminates because there is never a zero remainder. There is a repeating pattern: the 45 repeats infinitely.

$$
\begin{array}{r}
0.4545\ldots \\
11\overline{)5.0000} \\
\underline{-44} \\
60 \\
\underline{-55} \\
50 \\
\underline{-44} \\
60 \\
\underline{-55} \\
5
\end{array}
$$

The repeating pattern is often indicated with a line over the top of the block that repeats.

Example

These numbers are examples of repeating decimals:

$\frac{7}{11} = 0.636363\ldots = 0.\overline{63}$ $\qquad$ $\frac{7}{12} = 0.5833333\ldots = 0.58\overline{3}$

Converting a Decimal to a Fraction

A terminating decimal, such as 0.375, can be converted to a fraction. The place value of the last digit is the decimal is the denominator of the fraction.

Example

.375 can be written as $\frac{375}{1000}$.

$\frac{375}{1000}$ can be reduced to $\frac{375}{1000} = \frac{75 \bullet 5}{200 \bullet 5} = \frac{75}{200} = \frac{3 \bullet 25}{8 \bullet 25} = \frac{3}{8}$.

Fraction or Decimal?

Fractions are a useful way of representing numbers. They are particularly useful in expressing decimals that do not terminate.

Example

The fraction $\frac{1}{12}$ is an exact and compact way of representing the number $1 \div 12$.

$1 \div 12 = 0.08333\ldots$ The repeating decimal representation $0.08333\ldots$ is not compact because it does not terminate.

Comparing the relative size of two numbers may be easier if they are in decimal form rather than in fraction form.

Example

Deciding which is the smaller number, $\frac{5}{7}$ or $\frac{7}{12}$ is not an easy question when the numbers are given in fraction form.

But in decimal form, the question becomes an easier one.

$\frac{5}{7} = 0.7142857$ and $\frac{7}{12} = 0.58333\ldots$ $\quad 0.7142857 > 0.58333\ldots$

Zeros in Decimal Numbers

A zero to the left of a number does not change the value of the number.

Example

- 05.45 is the same number as 5.45.
- 0.45 is the same number as 0.45.

A zero in the middle of a number does change the value of the number.

Example

- In 0.305, the 5 is in the thousandths place, with a value of $\frac{5}{1000}$.
- In 0.35, the 5 is in the hundredths place, with a value of $\frac{5}{100}$.
- Thus, $0.35 > 0.305$, even though 0.35 has fewer digits then 0.305.

A zero to the right of a decimal point might make the number more precise.

Example

- In numbers, a zero at the end does *not* change the value.
  8 is the same number as 8.0.
- In measurements, a zero at the end indicates greater precision. 8.0 meters indicates the measurement was made to the nearest tenth of a meter, while 8 meters indicates the measurement was made only to the nearest meter.

Zeros in Measurement

Suppose you want to measure an athlete's long-jump distance. You could pace out the distance to the nearest meter, or you could measure the distance with a tape measure to the nearest tenth of a meter.

	Using 1-meter Paces	Using a Tape Measure
Measured Distance	8 meters	8.0 meters
Precision	Closer to 8 than to 7 or 9	Closer to 8.0 m than to 7.9 m or 8.1 m
Actual Distance	Between 7.5 and 8.5 m	Between 7.95 m and 8.05 m

In meter paces, the distance a certain athlete can jump is measured as 8 meters. Since no measurement is exact, the actual distance is between 7.5 meters and 8.5 meters.

Using a tape measure, the distance this athlete can jump is measured as 8.0 meters. The actual distance is between 7.95 meters and 8.05 meters.

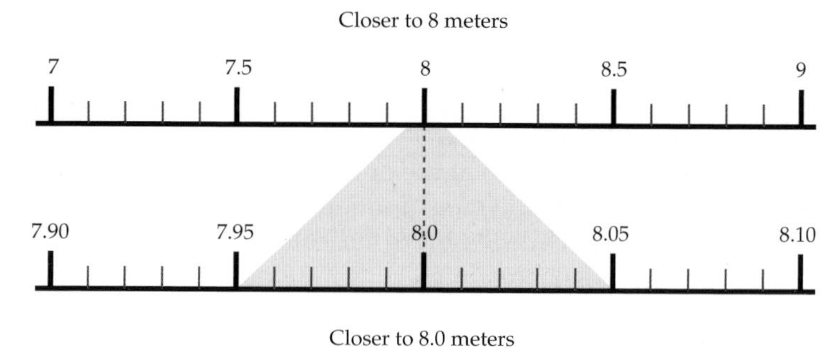

The measurement made using a tape measure is more precise than the one measured in 1-meter paces.

Adding and Subtracting Decimals

To add or subtract decimals, line up the digits according to their value. You also must answer this crucial question: "Where should I place the decimal point?"

There are three important steps when adding or subtracting decimals:

- Line up the digits according to place value.
- Line up the decimal points.
- "Carry" regrouped numbers from one column to the next, just as you do when adding or subtracting whole numbers.

Sometimes, both numbers have the same number of decimal places.

Example

$$\begin{array}{r} {\scriptstyle 1\quad\ \ } \\ 23.84 \\ +\ 4.94 \\ \hline 28.78 \end{array} \qquad\qquad \begin{array}{r} {\scriptstyle 1\ 12\ \ 18\ \ } \\ \cancel{2}\cancel{3}.\cancel{8}4 \\ -\ 4.94 \\ \hline 18.90 \end{array}$$

Note that the sum of 8 and 9 in the tenths column is 17. The 17 must be regrouped as 1 one and 7 tenths.

One of the 23 units must be regrouped as 10 tenths to give 18 – 9 = 9 in the tenths column.

Sometimes, the two numbers have different numbers of decimal places. By adding zeros at the end of one of the numbers, you can get both numbers to have the same number of decimal places.

Example

If you have 23.84 and 0.494, you can write the 23.84 as 23.840. Adding the zero will help you line up the digits but will not change the value.

$$\begin{array}{r} {\scriptstyle 1\ \ \ 1\qquad\ \ } \\ 23.840 \\ +\ 0.494 \\ \hline 24.334 \end{array} \qquad\qquad \begin{array}{r} {\scriptstyle 7\ 13\ 10} \\ 23.\cancel{8}\cancel{4}\cancel{0} \\ -\ 0.494 \\ \hline 23.346 \end{array}$$

Multiplying Decimals

You can multiply decimals using almost the same methods that you use to multiply whole numbers. The only extra step is to locate the position of the decimal point in the answer.

Example

Calculate: 49.32 • 6.85

Start with an estimate. For this example, a good estimate is 50 • 7 = 350.

Complete the multiplication, ignoring the decimal points. Do not worry about lining up the decimal points!

$$
\begin{array}{r}
\scriptstyle 5\ 1\ 1 \\
\scriptstyle 7\ 2\ 1 \\
\scriptstyle 4\ 1\ 1 \\
49.32 \\
\times \quad 6.85 \\
\hline
24660 \\
394560 \\
\underline{2959200} \\
3378420
\end{array}
$$

The result is 4,932 • 685 = 3,378,420—a number that has the same digits as the answer you want.

The important next step is to determine where to put the decimal point.

One way to do this is to use the estimate. Since the estimate was 350, the exact answer must be 337.8420.

Example

Another way to determine where to put the decimal point is to convert the decimals to fractions.

49.32 has 2 decimal places: $49.32 = \frac{4932}{100}$

6.85 has 2 decimal places: $6.85 = \frac{685}{100}$

$$49.32 \bullet 6.85 = \frac{4932}{100} \bullet \frac{685}{100} = \frac{4932 \bullet 685}{100 \bullet 100} = \frac{3,378,420}{10,000}$$

Divide 3,378,420 by 10,000.

The result has four digits to the right of the decimal point because there are four zeros in 10,000. The answer is 337.8420.

Check that this answer is close to the original estimate of 350. Yes, it is.

In general:

- Multiply decimals using the same methods that you use to multiply whole numbers. Ignore the decimal points and do not line up the decimal points.
- When you are done multiplying, count the total number of decimal places in each of the numbers multiplied, and put the decimal point in the product that many decimal places in from the right end.

Example

In 49.32 • 6.85, the number 49.32 has two decimal places and 6.85 has two decimal places. In the product the total number of decimal places you move to the left is four.

3,378,420 becomes 337.8420.

Dividing Decimals

You can divide a decimal by a whole-number divisor using the long division method.

Example

$43.446 \div 6$

The divisor is a whole number.

First, estimate: $43.446 \div 6 \approx 42 \div 6 = 7$

Now, use the long division method. Carefully line up the numbers so that the decimal point in the quotient is in the same position as the decimal point in the dividend.

The answer, 7.241, is close to the estimate of 7.

```
   7.241
6)43.446
  42
  ----
   14
   12
   ----
    24
    24
    ---
      6
      6
    ---
      0
```

Dividing a decimal by another decimal requires an extra step.

Example

$49.32 \div 6.85$

The divisor is a decimal.

Make the divisor a whole number by multiplying the dividend and divisor by 100.

$49.32 \div 6.85 = 4{,}932 \div 685$

Then use the standard method for division.

$49.32 \div 6.85 = 7.2.$

```
       007.2
685 )4932.0
    -4795
    -----
      137 0
     -137 0
     ------
          0
```

A general rule is:

Multiply the divisor and the dividend by the same power of 10 large enough to make each of them whole numbers. Then divide by the long division method.

Percent

Percent means "per hundred." One hundred percent equals "the whole."

Every percent can be rewritten as a fraction with a denominator of 100. Percents are easy to compare because all percents have the same denominator, 100.

Fractions with denominators of 100 can be represented as decimals, because the decimal system is based on multiplication and division by 10 and multiples of 10.

Example

$7.5\% = \frac{7.5}{100} = 0.075$

$75\% = \frac{75}{100} = 0.75$

$150\% = \frac{150}{100} = 1.5$

$100\% = \frac{100}{100} = 1$

CHAPTER 10

PERCENTS

Percents

A percent is part of 100. Because of this, it is easy to write a percent as a fraction with a denominator of 100.

Example

The fraction $\frac{25}{100}$ is written as 25%.

The number $\frac{100}{100}$ is written as 100%.

The fraction $\frac{125}{100}$ is written as 125%.

When writing a percent, use the percent sign, %.

Percents and Decimals

It is also easy to write a percent as a decimal in hundredths:

$1\% = 0.01 \qquad 10\% = 0.10 \qquad 150\% = 1.50$

A number even smaller than 1% is 0.5%, which as a decimal is written:

$$0.5\% = \frac{0.5}{100} = 0.005$$

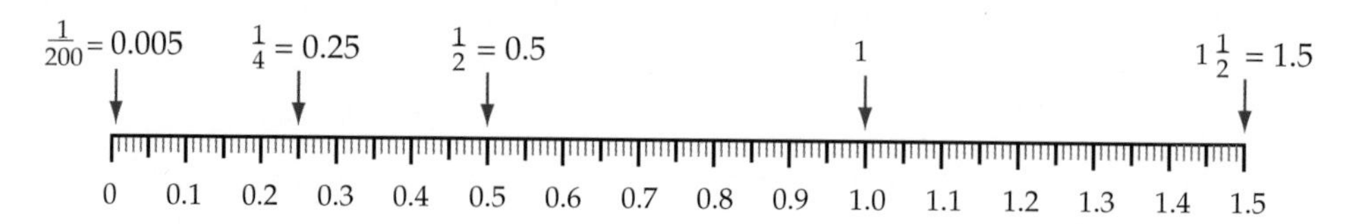

Percents and Fractions

Some fractions can easily be expressed as a percent.

Example

The fraction $\frac{7}{20}$ is easy to write as a percent, since 20 is a factor of 100:

$$\frac{7}{20} = \frac{7 \bullet 5}{20 \bullet 5} = \frac{35}{100} = 35\%$$

Other fractions can be expressed as a percent but take a little more work.

Example

$\frac{5}{8}$ 8 is not a factor of 100, but 8 • 125 = 1,000, so

$$\frac{5}{8} = \frac{5 \bullet 125}{8 \bullet 125} = \frac{625}{1000} = \frac{62.5}{100} = 62.5\%$$

You can use this fact $\frac{5}{8} = 62.5\%$ to help convert $\frac{5}{16}$ or $\frac{5}{24}$ into percents.

16 and 24 are multiples of 8.

$\frac{5}{16}$ $\frac{5}{16} = \frac{5}{8} \div 2 = 62.5\% \div 2 = 31.25\%$

$\frac{5}{24}$ $\frac{5}{24} = \frac{5}{8} \div 3 = 62.5\% \div 3 = 20.8\overline{3}\%$

Example

$\frac{20}{7}$ 7 is not a factor of 100, so the fraction needs to be divided out and expressed as a decimal before it is expressed as a percent.

$$\frac{20}{7} = 2.\overline{857142} \approx 2.\overline{857142} \bullet \frac{100}{100} \approx \frac{285.7142}{100} \approx 285.7\%$$

Percents Are Used to Compare Values

Here is some "free throw" data for three basketball players:

Player	Free Throws Attempted	Free Throws Made
Shaquille	300	180
Tayshaun	400	200
Marc	500	230

Marc had the most free throws, but he also had the most attempts.
A better comparison is provided by calculating "free throw percentage," free throws made divided by free throws attempted.

Player	Free Throw Percentage
Shaquille	$\frac{180}{300} = \frac{60}{100} = 60\%$
Tayshaun	$\frac{200}{400} = \frac{1}{2} = 50\%$
Marc	$\frac{230}{500} = \frac{23}{50} = \frac{46}{100} = 46\%$

The comparison shows that Shaquille has the highest success rate of free throw per attempt—and Marc has the lowest.

Free throw percentage is one of many types of statistics for which a score of greater than 100% is impossible. However, percentage comparisons greater than 100% can occur when a current performance is compared with a previous performance.

Example

Look at the following financial data.

Company	Last Year's Profit	This Year's Profit
Boots, Inc.	\$400,000	\$1,000,000
Shoes, Inc.	\$800,000	\$600,000

For Boots, Inc.

$$\frac{\text{This year}}{\text{Last year}} = \frac{1,000,000}{400,000} = \frac{10}{4} = 2.5 = 250\%$$

This year's report can say that profits were 250% of profits for the previous year. That means profits for this year were 2.5 times greater than last year's profits.

For Shoes, Inc.

$$\frac{\text{This year}}{\text{Last year}} = \frac{600,000}{800,000} = \frac{3}{4} = 75\%$$

Unfortunately, this year's report will have to say that profits were only 75% of profits for the previous year—or "down by 25%," because the percentage decreased from the previous year.

Example

Suppose you scored 27 out of 30 points on Test A, and 37 out of 40 points on Test B.

You missed 3 points on each test, but does this mean you did equally well on each?

To compare your scores, convert them to percents.

$$\frac{27}{30} = 90\% \qquad \frac{37}{40} = 92.5\%$$

Even though you missed the same number of points on each test, you scored higher on Test B.

Percent of an Amount

Many problems use multiplication to calculate a percentage of a given amount.

Example

A fundraiser claims that at least 90% of funds raised are donated to charity while no more than 10% is spent on administration.
This year they spent $150,000 of funds raised on administration and gave $1,100,000 to charity. Did they meet their 90% goal?

Solution:

Total funds raised were $1,100,000 + $150,000 = $1,250,000

$$90\% \text{ of } \$1{,}250{,}000 = \frac{90}{100} \bullet \$1{,}250{,}000 = 0.9 \bullet \$1{,}250{,}000$$
$$= \$1{,}125{,}000$$

Conclusion:

The fundraiser was $25,000 short of their target for the amount donated. They should have spent only $125,000 on administration and donated $1,125,000 to charity.

Percents Describe Changes in Value

Sometimes you need to figure a percent decrease.

You must have seen signs like this one on your trips through a shopping mall.

But what you really want to know is how much you need to pay—not how much the shop claims that you "save."

Example

The relationship between the prices is shown in the following diagram.

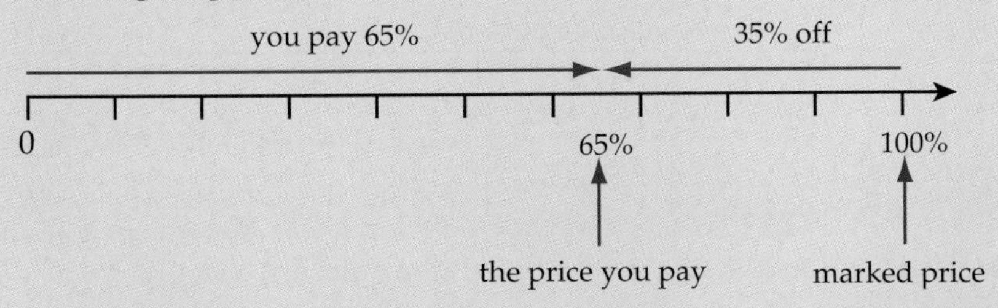

Marked price – Amount off = The price you pay

The marked price shown on the ball, \$49.95, is the 100% price.

The "amount off" is 35% of the marked price.

Thus, the price you would pay is 100% – 35% = 65% of the marked price.

Knowing this, you can calculate the price you would pay without having to first calculate the amount "saved."

$$\text{Sale price} = 65\% \text{ of } \$49.95 = \frac{65}{100} \bullet \$49.95 = 0.65 \bullet \$49.95$$

$$= \$32.4675$$

$$\approx \$32.47$$

Sometimes you need to find a percent increase.

Example

A restaurant bill for a birthday celebration comes initially to $54.60, but the small print at the bottom says

"10% tax to be added to all purchases."

Your final bill = initial amount + 10% of initial amount

To check the calculation of the final bill, you have two choices:

- Calculate 10% of $54.60 and add the result to $54.60
- Calculate 110% of $54.60

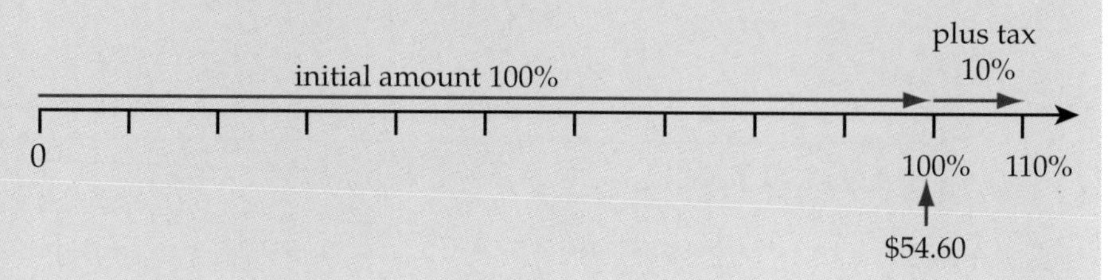

By mental calculation, the first method might be quicker:

Example

Mental calculation

Your final bill = $54.60 + $5.46 = $60.06

With a calculator, the second method would be quicker:

Example

Using a calculator

$$\text{Your final bill} = 110\% \text{ of } \$54.60 = \frac{110}{100} \bullet \$54.60$$

$$= 1.10 \bullet \$54.60$$

$$= \$60.06$$

If you discount by a certain percentage, like 30 percent, the discounted price is 70% of the original. If you add 30% to the discounted price, you do not get back to the original price.

Example

If you wanted to buy something that was $100 and it was at a 30% discount, the sales price would be $70.

If you start with the $70 and increase it 30%, you would get

$$\$70 + \$70 \bullet 0.30 = \$70 + 21 = \$91$$

When you discount the original price you multiplied by 70 % or $\frac{70}{100}$. To get back the original price, you must multiply the discounted price, in the case above, by $\frac{100}{70}$ which is about 1.4286. Increasing by 30%, on the other hand, would mean multiplying by 1.3, which is less than 1.4286. This can be surprising at first: you decrease by 30 percent but then must increase by about 43 percent to get back up to where you started.

In some comparison problems, the percentages can be greater than 100%.

Example

If a school grows in size from last year's population of 400 students to 1,000 students this year, by what percent has it increased?

The amount of increase is 1,000 – 400 = 600 students.
The percent of increase is:

$$\frac{\text{this year} - \text{last year}}{\text{last year}} = \frac{600}{400} = 1.5 = 150\%$$

You can say that the school population has increased by 150%.

Combining Percents

Some problems involve adding percents and others involve multiplying percents.

Adding Percents

Example

The initial restaurant bill could still be \$54.60, but you have decided to add a 15% tip.

Together with the 10% tax you end up paying an extra 25% of \$54.60.

$$\text{The amount you pay} = 125\% \text{ of } \$54.60$$
$$= \frac{125}{100} \bullet \$54.60$$
$$= \frac{5}{4} \bullet \$54.60$$
$$= \$68.25$$

Multiplying Percents

Some problems involve calculation of a percent of a previous answer that was also calculated as a percent.

Example

You might split the payment of the restaurant bill with three other family members, with your amount 25% of the total.

$$\text{The amount you pay} = \frac{25}{100} \bullet \frac{125}{100} \bullet \$54.60$$
$$= \frac{31.25}{100} \bullet \$54.60$$
$$\approx \$17.06, \text{ or } 31.25\% \text{ of the original amount.}$$

Example

At the end of each year the value of an investment increases by 10% of what it was at the start of the year. If the amount invested was \$2,500, what would its value be two years later?

Solution:

By the end of the first year the value will be 110% of \$2,500.

$$\frac{110}{100} \cdot \$2{,}500 = \$2{,}750$$

By the end of the second year, the value will be:

110% of the value at the end of the first year

= 110% of \$2,750

$$= \frac{110}{100} \cdot \$2{,}750 = \$3{,}025$$

Another way to solve this problem would be to multiply the percents directly:

$110\% \cdot 110\% = 121\%$

Then calculate 121% of the original amount:

$121\% \cdot \$2{,}500 = \$3{,}025$

POSITIVE AND NEGATIVE NUMBERS

CHAPTER 11

Positive Numbers, Zero, and Negative Numbers

Negative numbers are needed in situations such as these:

- $7 - 9 = ?$ The result is negative 2, which you can write as -2.
- Dwayne gained 7 pounds and then lost 9 pounds. How much did Dwayne gain or lose? The answer would be that he lost 2 pounds (–2 lbs). If his starting weight was w, his new weight is $w + 7 - 9 = w - 2$.
- A temperature that is below zero.

Positive numbers are numbers that are greater than zero.

Example

5,000, 4.75, 2, and 0.1

You say: positive 5,000, positive 4.75, positive 2, and positive 0.1

You write: 5,000 or +5,000, 4.75 or +4.75, 2 or +2, and 0.1 or +0.1

Negative numbers are numbers that are less than zero.

Example

$-5{,}000, -4.75, -0.1, -\frac{1}{2}$

You say: negative 5,000, negative 4.75, negative 0.1, negative one-half

You write: $-5{,}000, -4.75, -0.1, -\frac{1}{2}$

The number zero is neither positive nor negative.

When the symbols "+" and "–" are used before a single number to indicate whether it is positive or negative, the "+" is called a *positive* sign and the "–" is called a *negative* sign. The same symbols, "+" and "– ," are also used for the operations *addition* and *subtraction*. Even though the same symbols are used, the difference between the operation of *addition* and the sign of a *positive* number is important. Similarly, the difference between the operation of *subtraction* and the sign of a *negative* number is important. To avoid confusion, it is best to use the words plus and minus to refer to the operations, but not to use the words *plus* and *minus* to refer to positive and negative numbers.

Example

- Read $7 - 9 = -2$ as, "positive seven minus positive nine equals negative 2."
- Read $7 - (-9) = 16$ as, "positive seven minus negative nine equals positive 16."

Positive and Negative Numbers on the Number Line

To show negative numbers on the horizontal number line, the number line must be extended to the left beyond zero. If the number line is drawn horizontally, then every positive number can be marked as a point on the right side of 0, and every negative number can be marked as a point on the left side of 0.

The diagram below shows a number line with four positive numbers marked to the right of 0 and three negative numbers marked to the left of 0.

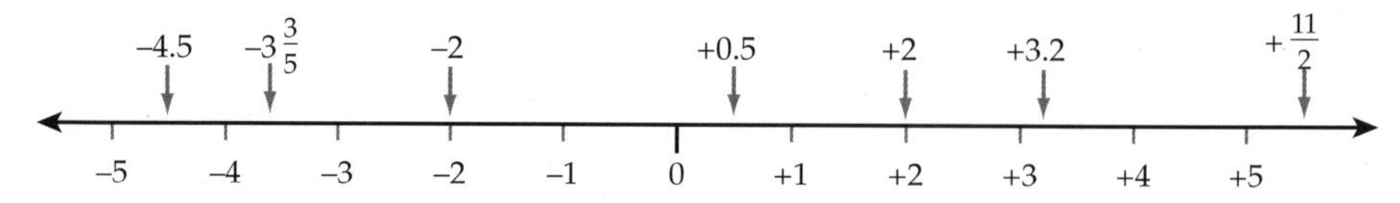

Sometimes, the number line is drawn vertically, with positive numbers above 0 and negative numbers below 0.

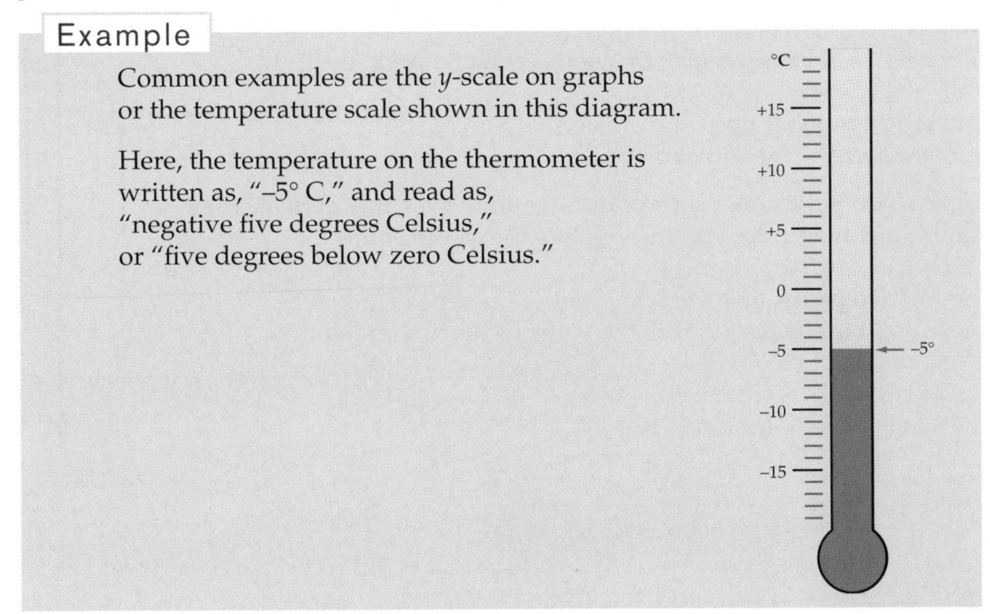

Example

Common examples are the y-scale on graphs or the temperature scale shown in this diagram.

Here, the temperature on the thermometer is written as, "–5° C," and read as, "negative five degrees Celsius," or "five degrees below zero Celsius."

Greater Than and Less Than: Putting Numbers in Order

On horizontal number lines, the direction to the right is called the *positive direction.* The direction to the left is called the *negative direction.*

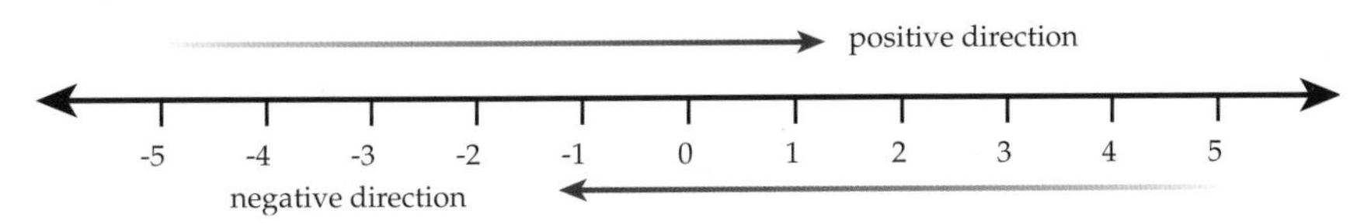

For any two points on a horizontal number line, the value of the point on the right is always *greater than* the value of the point on the left.

For any two points on a horizontal number line, the value of the point on the left is always *less than* the value of the point on the right.

On vertical number lines, the direction up is called the *positive direction.* The direction down is called the *negative direction.*

For any two points on a vertical number line, the value of the higher point is always *greater than* the value of the lower point.

For any two points on a vertical number line, the value of the lower point is always *less than* the value of the higher point.

The symbol for "*is greater than*" is $>$.
The symbol for "*is less than*" is $<$.

Example

- The point 4.5 is above or to the right of the point 2.3, so $4.5 > 2.3$.
- The point -4 is below or to the left of the point -1, so $-4 < -1$.
- The point 20 is above or to the right of the point -30, so $20 > -30$.

You can also make comparisons when you have three or more numbers.

Example

If you have -3, 0, and 5, the number -3 is less than 0, and 5 is greater than 0. You can express these individually as $-3 < 0$ and $5 > 0$, and you can also express these together.

- In increasing order: $-3 < 0 < 5$
- In decreasing order: $5 > 0 > -3$

The symbol for *"is greater than or equal to"* is $\geq$.
The symbol for *"is less than or equal to"* is $\leq$.

The statement $x \leq 3$ is true if x represents 3, because 3 = 3, or it is true if x is any number less than 3.

The statement $x \geq -2$ is true if x represents –2, because –2 = –2, or it is true if x is any number greater than –2.

Absolute Value

The *distance* between 0 and a number on a number line is the *absolute value* of that number. For all numbers other than 0, this distance is a positive number.

The symbol, "$| \;|$" is used to represent absolute value, as in $|x|$

The absolute value of +5 is 5: $|+5| = +5$

The absolute value of –5 is 5: $|-5| = +5$

Example

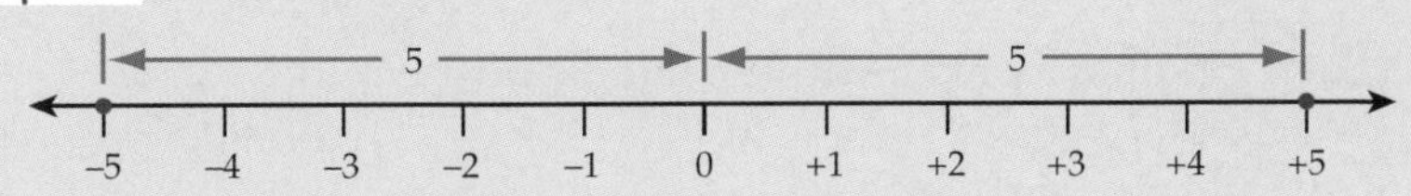

These two numbers, –5 and +5, have the same *absolute value.*
So, $|-5| = |+5|$, which means –5 and +5 are the same distance from 0 on the number line.

Each is called the *opposite* of the other.

–5 is the opposite of +5. +5 is the opposite of –5.

Adding Positive and Negative Numbers

You can use the number line to show the addition $a + b$, where a and b are any two numbers (positive, negative, or 0), as follows:

- Start at a (which may be positive, negative, or 0).
- From a, move a distance $|b|$ to the right if b is positive or to the left if b is negative.
- You will end up at $a + b$.

You are already familiar with adding two positive numbers.

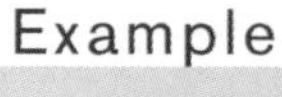

Example

$5 + 8$

Start at 5. Go a distance of 8 in the positive direction.

You end up at 13. $5 + 8 = 13$

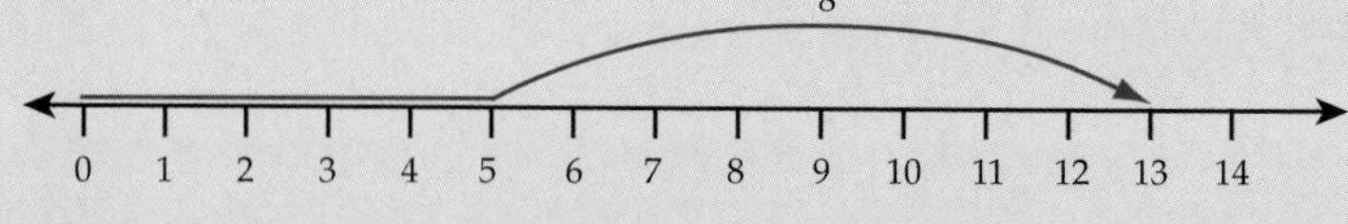

Say, "Positive 5 plus positive 8 equals positive 13."

You can add a positive number to a negative number.

Example

$-5 + 8$

Start at –5. Go a distance of 8 in the positive direction.

You end up at 3. $-5 + 8 = 3$

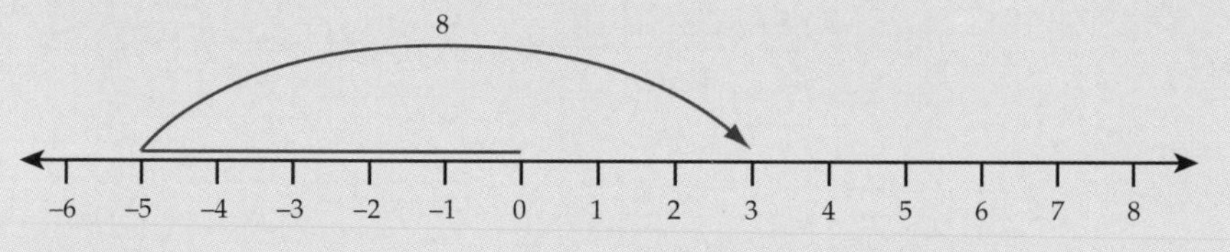

Say, "Negative 5 plus positive 8 equals positive 3."

You can add a negative number to a positive number.

Example

$5 + (-8)$

Start at 5. Then, because –8 is negative, go a distance of 8 in the negative direction. You end up at –3. $5 + (-8) = -3$

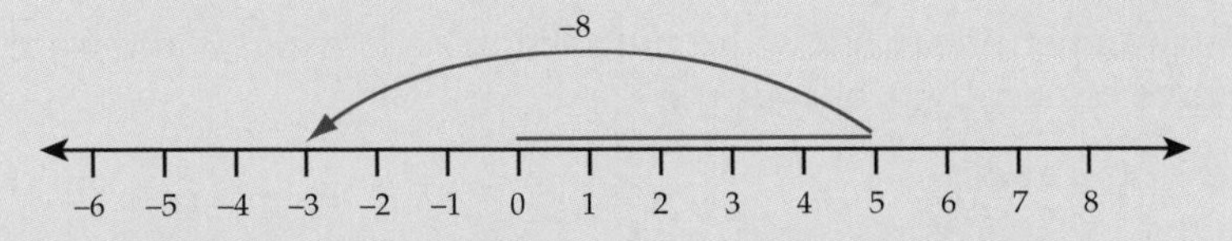

Say, "Positive 5 plus negative 8 equals negative 3."

You can add a negative number to a negative number.

Example

$-5 + (-8)$

Start at –5. Then, because –8 is negative, go a distance of 8 in the negative direction. You end up at –13. $-5 + (-8) = -13$

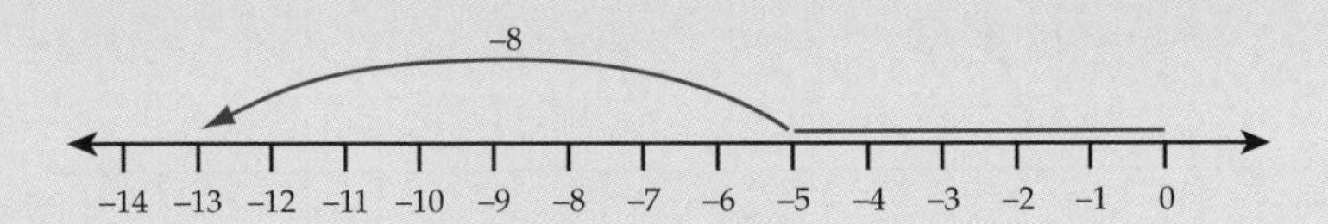

Say, "Negative 5 plus negative 8 equals negative 13."

Subtracting Negative and Positive Numbers

You can use a number line to show the subtraction, $a - b$.

Subtraction is the difference between two numbers. This difference can be represented using the line segment between the two numbers on the number line. This is true whether the numbers are positive or negative.

You can use the number line to show the subtraction $a - b$, where a and b are any two numbers (positive or negative), as follows:

- Make points a and b.
- Find the distance between a and b.
- Determine the direction from point b to point a. If the direction is positive, the answer is positive; if the direction is negative, the answer is negative.

You can subtract a positive number from a positive number.

Example

5 – 3

Mark +5 and +3 on the number line. The distance from 5 to 3 is 2. the direction from +3 to +5 is a positive direction. 5 – 3 = +2

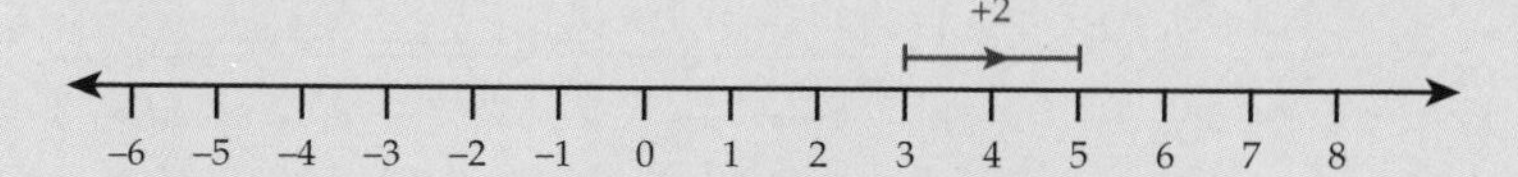

Say, "Positive 5 minus positive 3 equals positive 2."

Note that 5 – 3 = 2 and 5 = 3 + 2 look the same on the number line. In fact, adding 3 to both sides of the first equation gives the second equation.

You can subtract a positive number from a negative number.

Example

$-1 - 5$

Mark –1 and +5 on the number line. The distance from –1 to +5 is 6.
The direction from +5 to –1 is a negative direction. $-1 - 5 = -6$

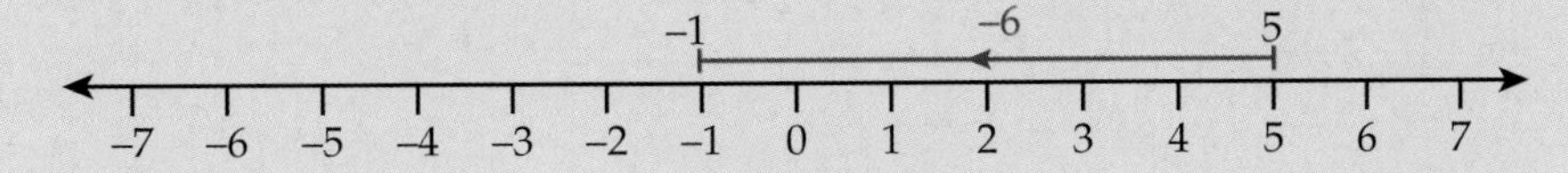

Say, "Negative 1 minus positive 5 equals negative 6."

You can subtract a negative number from a negative number.

Example

$-5 - (-3) = -2$

Mark –5 and –3 on the number line. The distance from –5 to –3 is 2.
The direction from –3 to –5 is a negative direction. $-5 - (-3) = -2$

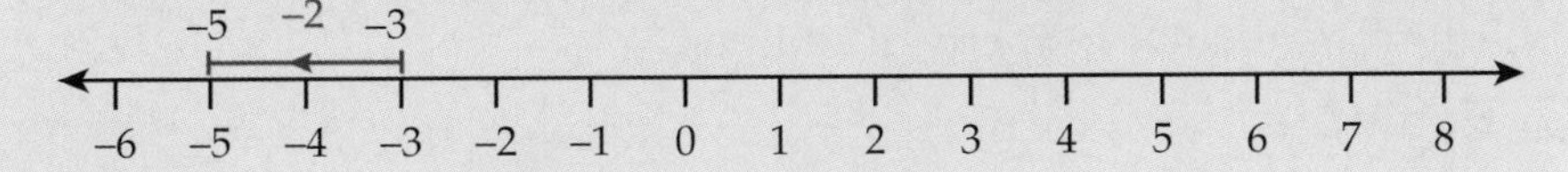

Say, "Negative 5 minus negative 3 equals negative 2."

You can subtract a negative number from a positive number.

Example

$1 - (-5) = 6$

Mark +1 and –5 on number line. The distance from 1 to –5 is 6.
the direction from –5 to +1 is a positive direction. $1 - (-5) = 6$

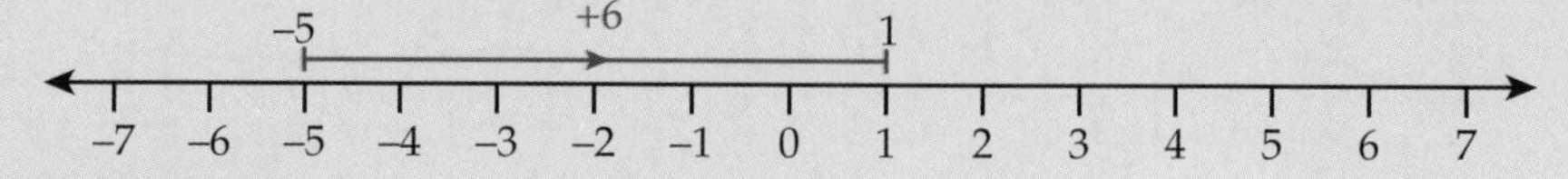

Say, "Positive 1 minus negative 5 equals positive 6."

Note that in these examples, the number line diagram of $x - y = z$ would be the same as the number line diagram of $x = y + z$.

Number Properties of Addition and Subtraction

The Commutative Property of Addition

$a + b = b + a$ is true for all numbers, a and b.
This is the commutative property of addition.

Example

$$-5 + 2 = -3$$
$$2 + (-5) = -3$$
$$\text{so } -5 + 2 = 2 + (-5)$$

However, the commutative property does not apply to subtraction. In general, $a - b$ does not equal $b - a$.

Example

$-5 - 2 = -7$, but $2 - (-5) = 7$

These two equations are different, showing that the commutative property does not apply to the operation of subtraction.

The Associative Property of Addition

$(a + b) + c = a + (b + c)$ is true for all numbers a, b, and c.
This is the associative property of addition. This property allows you to add a group of numbers in any order.

Example

Consider $5 + (-3) + (-6)$ and $5 + [-3 + (-6)]$:

You can add the first two numbers, and then add that sum to the third number.	You can add the second and third numbers, and then add that sum to the first number.
$[5 + (-3)] + (-6) = 2 + (-6) = -4$	$5 + [-3 + (-6)] = 5 + (-9) = -4$

In each case, the result is the same, illustrating the associative property of addition.

However, the associative property does not apply to subtraction. In general, $(a - b) - c \neq a - (b - c)$.

Example

Consider $[5 - (-3)] - (-6)$ and $5 - [-3 - (-6)]$

$[5 - (-3)] - (-6) = 8 - (-6) = 14$

$5 - [-3 - (-6)] = 5 - 3 = 2$

The two results are different, showing that the associative property does not apply to the operation of subtraction.

Replacing a Subtraction with an Addition

A useful rule is $a - b = a + (-b)$. This means that you get the same result when *subtracting* b as you do when *adding* its opposite, $-b$.

Example

$6 - 4 = 6 + (-4) = 2$ $\quad$ $-12 - 5 = (-12) + (-5) = -17$

Multiplying Positive and Negative Numbers

Multiplying a Positive Number by a Positive Number

When multiplying a positive number by a positive number, the result is positive.

Example

$4 \bullet 4 = 16$ and $8 \bullet 3 = 24$

Multiplying a Positive Number by a Negative Number

When multiplying a positive number by a negative number, the result is negative.

Example

The expression $3 \bullet -5$ can be thought of as 3 groups of negative 5:

$3 \bullet (-5) = (-5) + (-5) + (-5) = -15$ (a negative number)

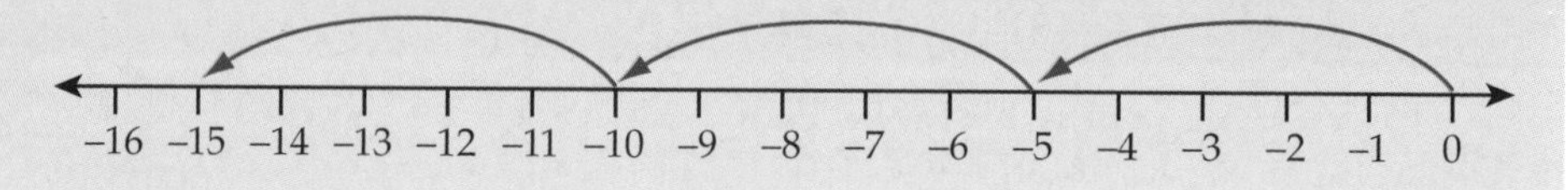

Multiplying a Negative Number by a Positive Number

When multiplying a negative number by a positive number, the result is negative.

Example

Since multiplication is commutative, the expression $(-5) \bullet 3$ is the same as $3 \bullet (-5)$. Using the previous example, the result is -15.

Multiplying a Negative Number by a Negative Number

When multiplying a negative number by a negative number, the result is positive.

Example

The expression, $-3 \bullet (-5)$, can be thought of as the opposite of $3 \bullet (-5)$. Since this equals -15, using the previous examples you see that $-3 \bullet (-5) = 15$ (a positive number).

Summary

These examples illustrate four simple rules:

- A positive number multiplied by a positive number is a positive number.
- A positive number multiplied by a negative number is a negative number.
- A negative number multiplied by a positive number is a negative number.
- A negative number multiplied by a negative number is a positive number.

More Number Properties

Now that you are considering negative as well as positive numbers, there are some more number properties that can be illustrated.

The Inverse Property of Addition

Every number has an *opposite*. Adding a number and its opposite results in 0. The opposite of a number is called the *additive inverse*.

Example

The opposite of -5 is $+5$, since $-5 + (+5) = 0$.

Thus, -5 and 5 are additive inverses of each other.

The Inverse Property of Multiplication

Every nonzero number has a *reciprocal*. Multiplying a nonzero number by its reciprocal gives 1. The reciprocal of a number is called the *multiplicative inverse*.

Example

The reciprocal of $\left(-\frac{5}{2}\right)$ is $\left(-\frac{2}{5}\right)$, since $\left(-\frac{5}{2}\right) \bullet \left(-\frac{2}{5}\right) = +1$.

Thus, $\left(-\frac{5}{2}\right)$ and $\left(-\frac{2}{5}\right)$ are multiplicative inverses of each other.

The Commutative Property of Multiplication

$a \bullet b = b \bullet a$, where a and b can be any numbers, including negative numbers.

The Associative Property of Multiplication

$(a \bullet b) \bullet c = a \bullet (b \bullet c)$, where a, b, and c can be any numbers, including negatives.

Applying the laws of signs for multiplying nonzero numbers:

- If just one of these three numbers is negative, then the product will be negative.
- If exactly two of these numbers are negative, then the product will be positive.
- If all three of these numbers are negative, then the product will be negative.

The Distributive Property

$a \bullet (b + c) = (a \bullet b) + (a \bullet c)$, where a, b, and c can be any numbers, including negative numbers.

Example

$a \bullet (b + c) = (a \bullet b) + (a \bullet c)$

$-5 \bullet [2 + (-3)] = (-5 \bullet 2) + [-5 \bullet (-3)]$

$-5 \bullet (-1) = -10 + 15$

$5 = 5$

CHAPTER 12

GRAPHS

Coordinate Plane

A coordinate plane allows you to represent zero-dimensional objects, one-dimensional objects, and two-dimensional objects.

1. A point is a zero-dimensional object.

   A point has no dimensions. It has no size. A point is just a location. A point is represented by a dot. The dot on the page does have size: length, width, and depth—otherwise you would not be able to see it. However, the point that this dot represents is just a mathematical object with no size.
   You can think of a point as a position or a location in space.

2. A line is a one-dimensional object.

   A line has only length. It has no width or depth. A line extends forever in both directions. A line is represented by a thin black bar. The line on the page has width and depth, as well as length—otherwise you would not be able to see it. However, the line that it represents is a mathematical object with no width or depth, just length. You can think of a line as the edge of something or the direction of something.

3. A plane is a two-dimensional object. A polygon is a two-dimensional object on a plane.

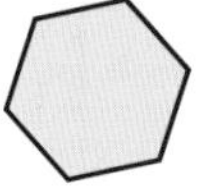 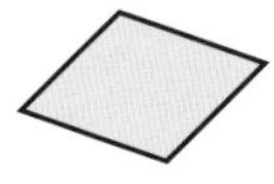 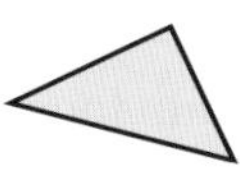 

   A polygon has area, since it has both length and width. But it has no volume, since it has no depth. You can think of a plane as a surface. Figures on a plane, such as polygons, are like areas on the surface.

Chapter 12

A special kind of plane used in mathematics is the *coordinate plane*, sometimes called the *Cartesian plane* after its inventor, René Descartes. It is one of the most useful tools in mathematics.

The coordinate plane makes it possible to represent relationships between numbers and variables geometrically as well as algebraically. It shows a way in which geometry corresponds to algebra.

To construct a coordinate plane, draw a horizontal number line. The number line locates numbers along a single line like a ruler. Make sure you have marked 0 on your line. To the left of 0 are the negative numbers and to the right are the positive numbers.

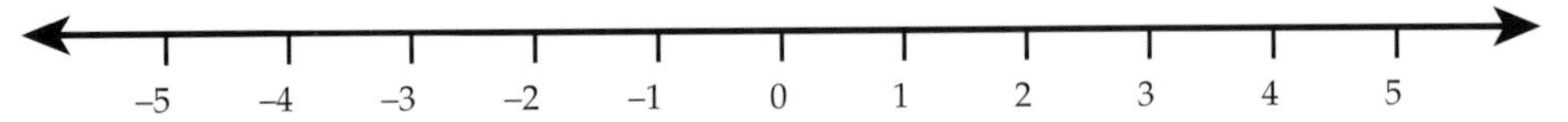

Next, draw a vertical number line that intersects the horizontal line at the 0 point. The vertical line and the horizontal line should be perpendicular (at right angles) to each other.

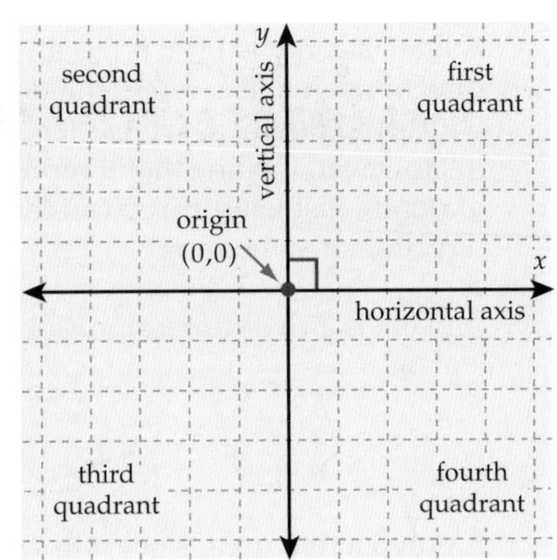

The horizontal line is called the x-axis. The vertical line is called the y-axis.

On the y-axis, the negative numbers are below 0, and the positive numbers are above 0. The x and y axes intersect at only one point. It is the point where $x = 0$ and $y = 0$. It is called the *origin* of the coordinate plane.

The two axes divide the page into four parts, called *quadrants*. In the first quadrant, x is positive, and y is positive.

Plotting Points on the Coordinate Plane

The coordinate plane has two dimensions. In this case, the dimensions are named the x-dimension and the y-dimension. Every point can be located by referring to its x-value and its y-value.

Every point (location) in the coordinate plane is identified by two numbers, the x-value and the y-value. A point is identified by giving these two values. By convention, the x-coordinate is always given first, followed by a comma, followed by the y-coordinate: (x, y). (x, y) is called an *ordered pair*.

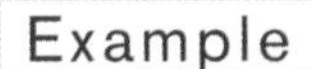

(3, 5) is the point located at $x = 3$ and $y = 5$.

3 is the x-coordinate and 5 is the y-coordinate.

(3, 5) is located in the first quadrant.

(–4, –2) is the point located at $x = -4$ and $y = -2$.

–4 is the x-coordinate and –2 is the y-coordinate.

(–4, –2) is located in the third quadrant.

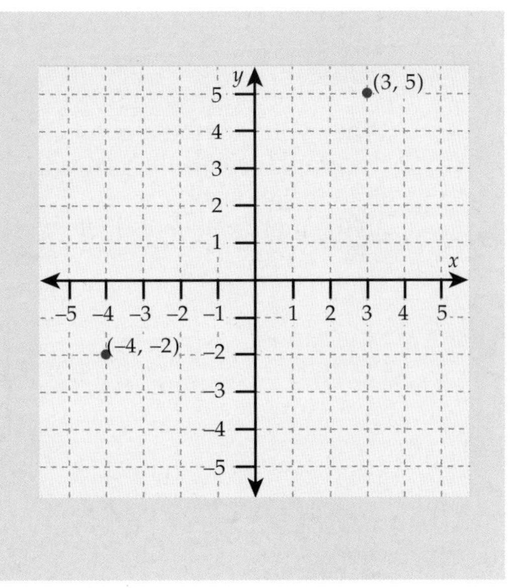

The graph of a point shows the relationship between its two coordinates. Since each point relates an x-value to a y-value, the most common use of coordinate graphs is showing the relationship between two quantities.

Plotting Points from a Table

If you have a table which relates two variables, those variables can be plotted on the coordinate plane.

Example

The following table gives corresponding values for variables x and y.

	A	B	C	D	E	F	G	H
x	+3	+4	+3	+2	–1	–6	–4	+2
y	–2	+1	+4	+5	+6	+1	–3	–3

When you plot these points, the graph reveals the pattern in the relationship: all the points lie on a circle.

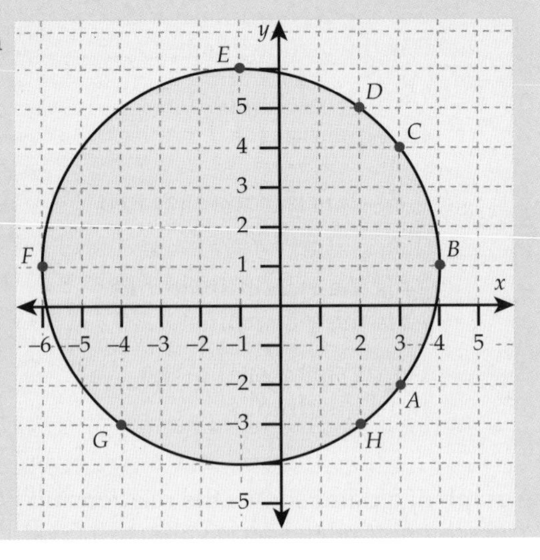

Sketching a Graph to Represent Situations

Graphs can represent situations.

Example

This graph shows a right triangle.

The x-value of the point P represents the length of the base, and the y-value represents the height of the triangle.

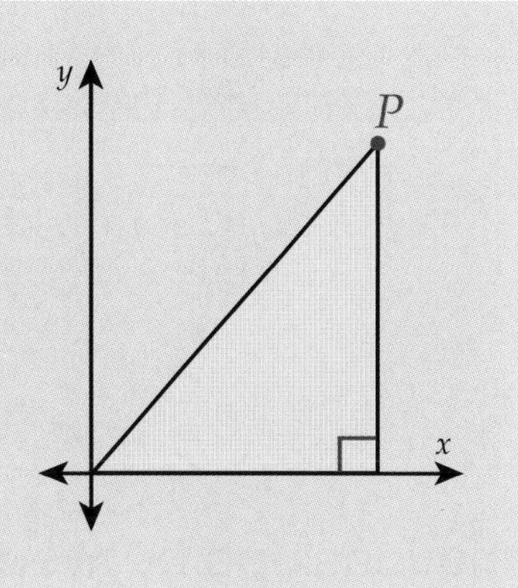

Example

The x-axis represents time in minutes and the y-axis represents distance in feet.

The line graph represents the water flowing at a rate of 5 feet per minute.

The water flows 10 feet in 2 minutes.

The water flows 20 feet in 4 minutes.

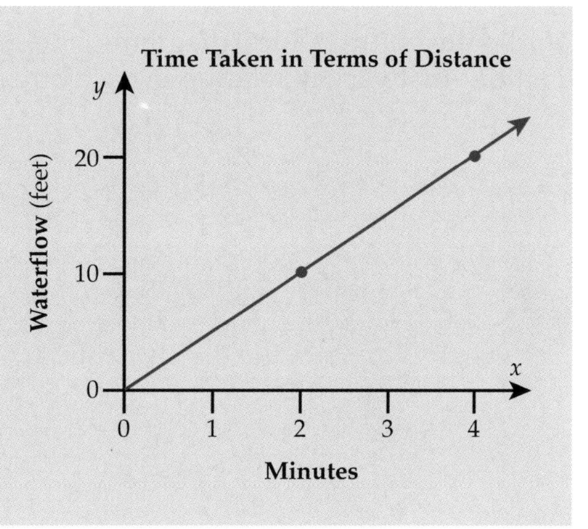

Example

The x-axis represents distance in miles, and the y-axis represents the total cost of the ride in dollars.

A cab ride of 5 miles costs $2.
A cab ride of 10 miles costs $4.

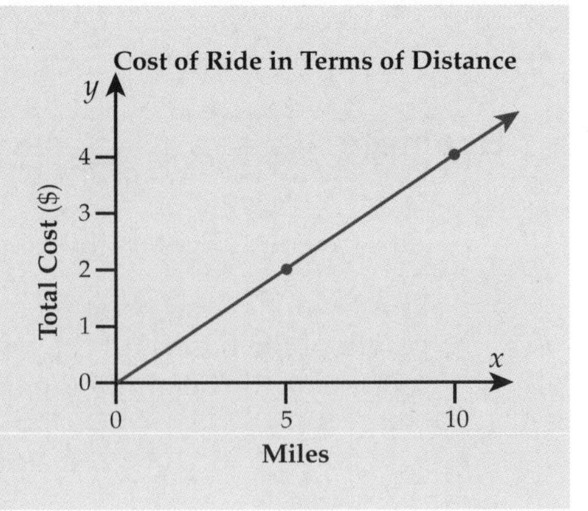

Relationships between Quantities

A *quantity* is an amount that can be counted or measured.

Coordinate graphs, tables, diagrams, and equations (or formulas) can represent quantities that vary in relation to each other.

Discrete Situations

Here is a typical situation where there are two quantities that vary (or change) in relation to each other:

Example

A large number of identical books are stacked in a school's book room. Each book is 3 inches thick. Show the height of the stack for any number of books.

Note: A way to solve this problem is detailed below.

Variables

A quantity that varies is called a *variable quantity* or *variable.*

In the situation above, there are two quantities that vary.
Quantity 1: The number of books
Quantity 2: The height of the stack

These two quantities vary in relation to each other because:

- As the number of books increases, the height of the stack increases.
- As the number of books decreases, the height of the stack decreases.

Let b = the number of books in the stack, and h = the height of the stack.

Diagrams

You can represent the situation with a diagram.
What is the height of the stack if there are 3 books?
If there are 5 books?

You can show the height of the stack.

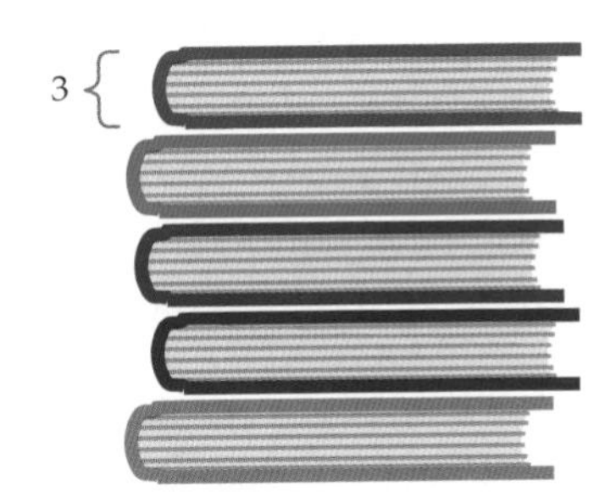

Tables

This table shows some pairs of values for the two quantities that vary in this situation:

Number of Books	1	2	3	5	10	12	b
Height of the Stack	3 in	6 in	9 in	15 in	30 in	36 in	$3b$

Equations (or Formulas)

Using b to stand for the number of books in the stack and using h to stand for the height, you can write an equation (or formula) that defines the relationship between these two quantities:

$$h = 3 \bullet b$$

or

$$h = 3b$$

Graphs

To plot a graph, you need to determine what scale to use.

- The set of x-values for the relationship is called the *domain*. The domain consists of the numbers you put into your equation. You could use 0 to 10 books as your domain.
- The set of y-values for the relationship is called the *range*. Given the domain you have chosen, the range is 0 through 30 inches.

If you have centimeter graph paper that is 16 centimeters wide, you could use each centimeter on the x-axis to stand for one book.

If the height of your graph paper is 16 centimeters then you could not have one centimeter for each inch of height because you could not plot some of the values. Instead you could have the y-scale be 3 inches for each centimeter on the y-axis.

A graph shows the number of books and the height of the stack as a single point.

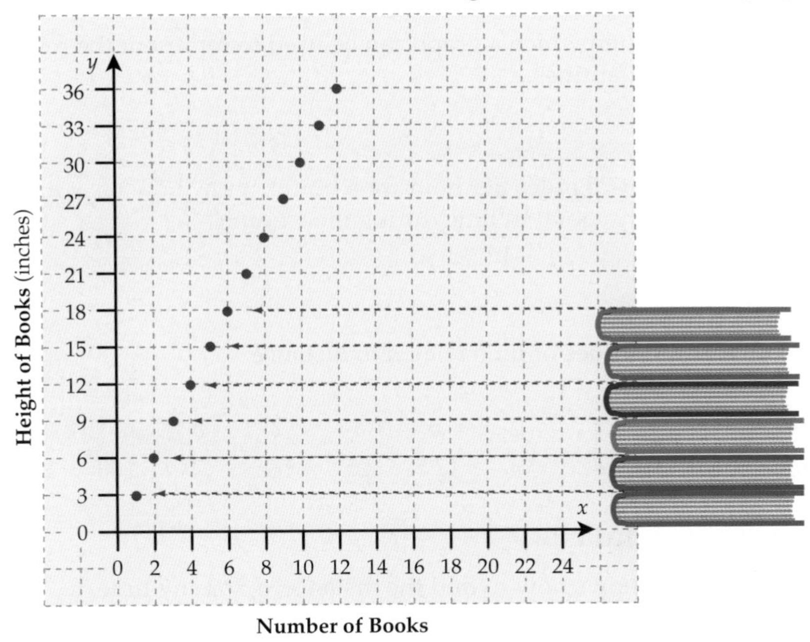

This is the graph of the formula, $h = 3b$, where b is the number of books and h is the height of the stack. For any point on the graph, read down to the place where it intersects the x-axis to find the x-value and read across to the place where it intersects the y-axis to find the y-value.

The graph shows the relationship between the number of books and the height of the stack. In this situation, the graph is not a solid line, but rather a set of discrete points. Why do you think this is?

The x-values represent the number of books in the stack. The number of books can only be a whole number; you cannot have a fraction of book.

This type of graph, with points rather than a solid line, is called *discrete*.

Continuous Situations

Here is another situation in which a graph can be used to represent the relationship between two variable quantities.

Example

A bird flew at 25 miles per hour for 6 hours. Show how far it had traveled at any point in time during the 6 hours.

This table shows some values of time related to distance:

Time in Hours	t	0	.5	1	2	3	4	5	6
Distance in Miles	d	0	12.5	25	50	75	100	125	150

Equations (or Formulas)

Here is a formula that shows how to find the distance, d, for any time, t:

$$d = 25\frac{\text{miles}}{\text{hour}} \bullet t$$

Graphs

A graph shows time traveled and distance flown as a single line.

In this graph, the line labeled $d = 25t$ shows all the points where $d = 25t$.
For any point in time on the t-axis (less than 6), you can go up to the line and read across to the d-axis to find the distance traveled by that time.

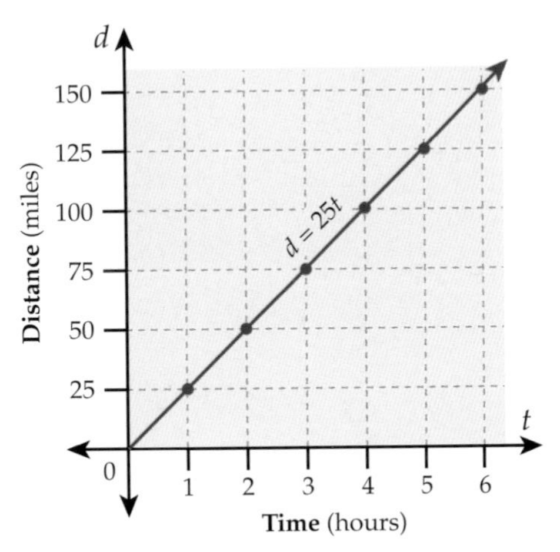

The t-axis shows time that is measured in hours, and the y-axis shows distance which is measured in miles.

The graph of the formula $d = 25t$ shows the relationship between time and distance as a line. The line is straight because the speed is a constant rate. If the bird accelerated, the line would curve up because the bird covers more distance in the same amount of time. What happens to the line if the bird slows down?

This graph is a straight line, not a series of points, as in the stack of books example. The line is solid because time can be any value along the t-axis. This type of graph is called *continuous.*

Summary

Any formula can be graphed. Formulas that show the relationship between two variables can be graphed using one variable as the horizontal axis and the other as the vertical axis.

The line of the graph shows the relationship between the two variables in the formula. When the axes represent quantities, the number values for each axis refer to the units of the quantity.

GEOMETRIC FIGURES

CHAPTER 13

Lines

A line is a straight one-dimensional object that extends forever in the two directions.

A line is uniquely determined by two distinct points. Lines are generally labeled with two letters corresponding to two points on the line.

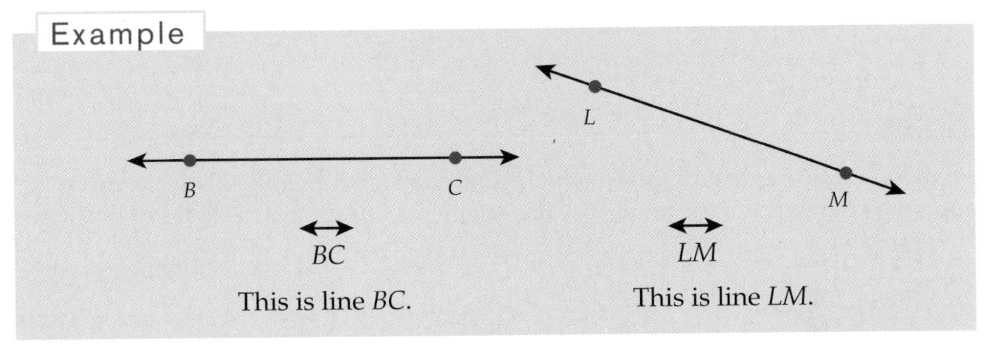

Rays

A ray is a part of a line that begins at a point and extends forever in one direction. Rays are labeled with two letters corresponding to the beginning point and another point on the ray.

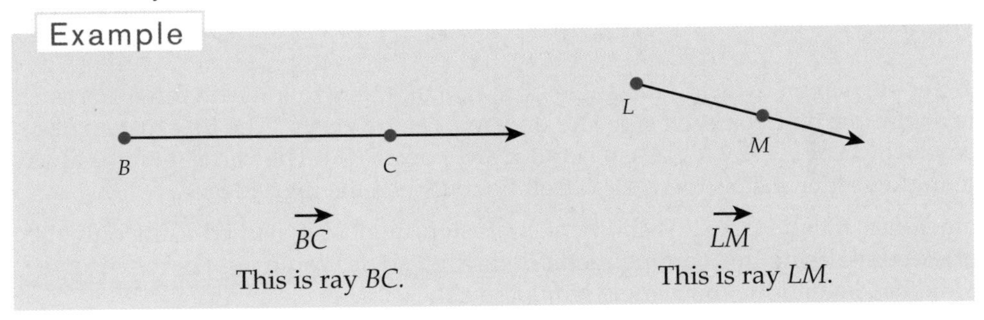

Chapter 13

Line Segments

A line segment is a part of a line between two points. Line segments are generally labeled with two letters corresponding to the two points.

Example

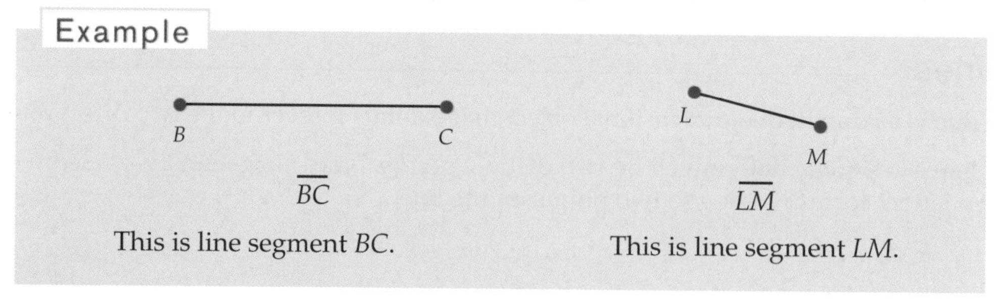

This is line segment BC.

This is line segment LM.

Angles

Two rays, or line segments, with a common endpoint form an angle. The common endpoint is the *vertex* of the angle; the line segments or rays are the sides of the angle.

Example

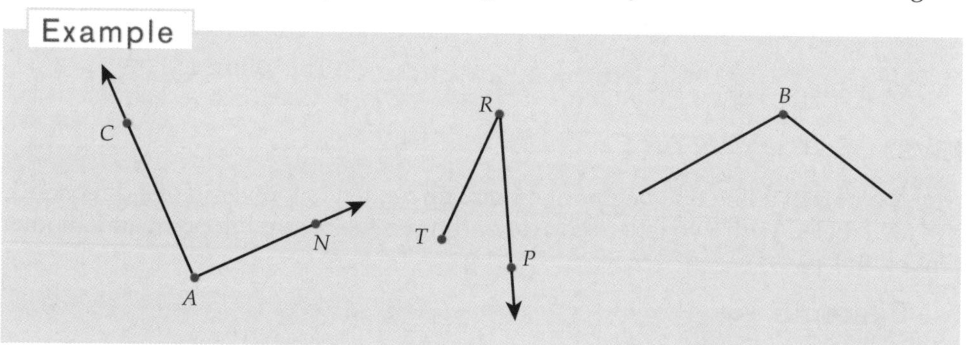

Angles are named by using the angle sign, ∠, and letters to indicate three points on the angle: one point on each side and one point at the vertex. The first two angles above would be $\angle CAN$ (or $\angle NAC$) and $\angle TRP$ (or $\angle PRT$). The vertex point is always the middle letter, just as the vertex itself lies between the two sides.

Sometimes, if there is no possibility of confusion, angles are named using only the vertex point. Using this naming method, the first angle would be $\angle A$, the second would be $\angle R$, and the third angle would be $\angle B$.

Measuring Angles

The *measure* of an angle is a number that indicates its size. The measure of an angle is often given in *degrees*. The symbol for degrees is °. Examples are 60° and 90°.

The table below shows angles in increasing size from 15° to 180°.

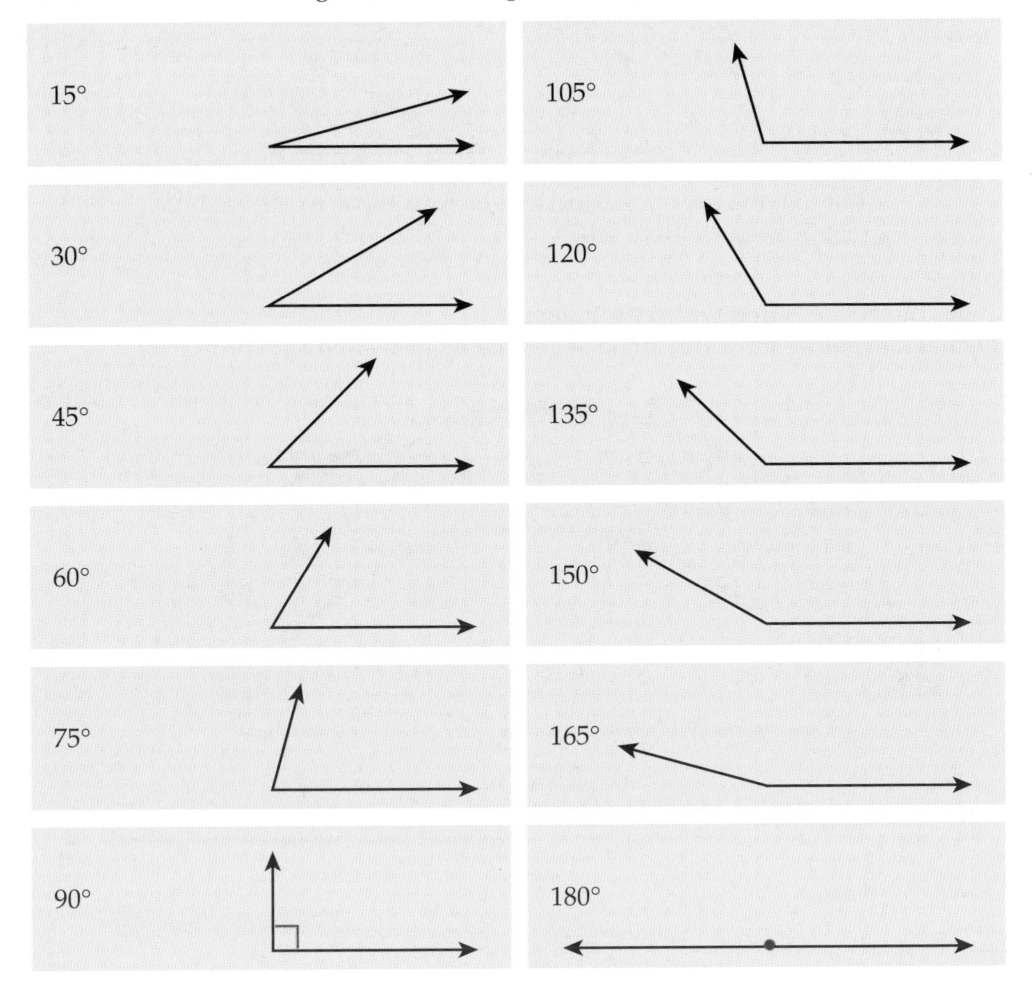

Note that the length of the sides shown on an angle has nothing to do with the angle's measure. The measure of an angle depends only on the amount of rotation from one side to the other side.

Example

$\angle A$ has a greater measure than $\angle B$.

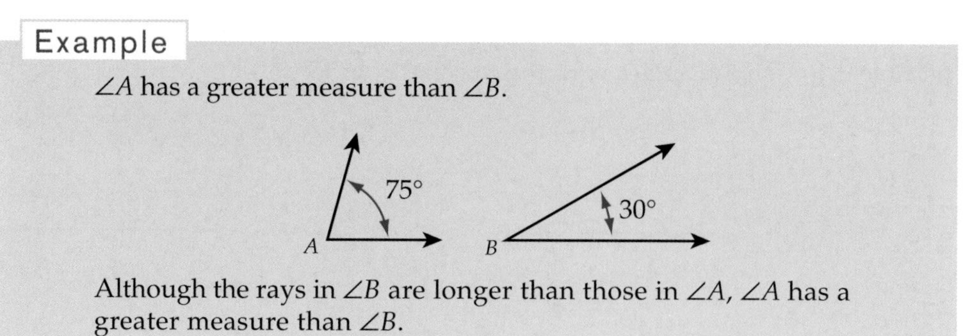

Although the rays in $\angle B$ are longer than those in $\angle A$, $\angle A$ has a greater measure than $\angle B$.

Angles can be measured with a protractor.
The diagram below shows an angle of 105° being measured by a protractor.

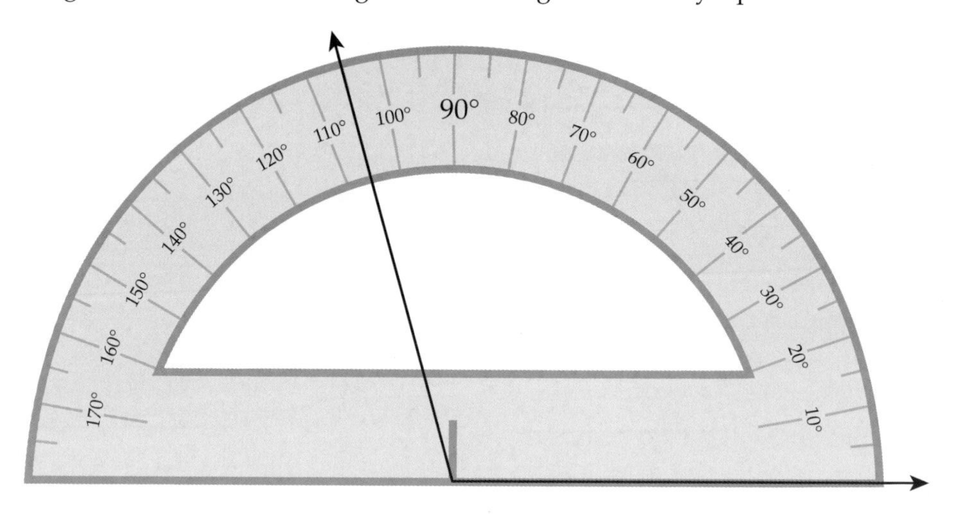

Congruent Angles

If the measures of two angles are equal then they are said to be *congruent angles*. You can show that angles are congruent with congruence marks: | and ||.

Example

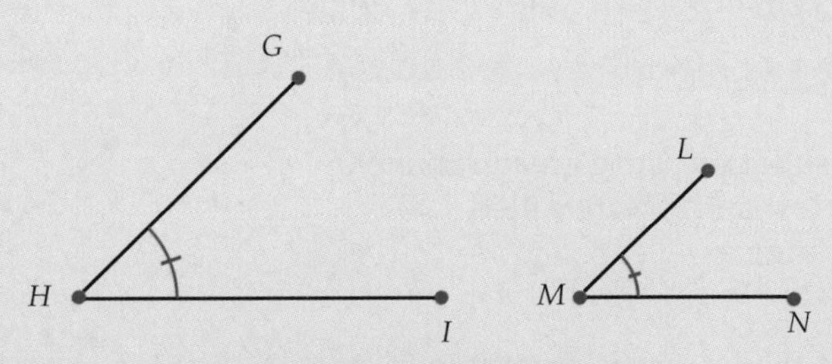

Although the line segments have different lengths, $\angle GHI$ and $\angle LMN$ have the same measure.

$\angle GHI = \angle LMN = 45°$.

$\angle GHI$ and $\angle LMN$ are congruent angles. Congruent angles have equal measures.

Defining Angle Types

Angles are classified according to their size. Different types of angles have different measures.

Angles that measure greater than 0° and less than 90° are called *acute* angles.

Angles that measure greater than 90° and less than 180° are called *obtuse* angles.

Angles that measure greater than 180° and less than 360° are called *reflex* angles.

A *right* angle is formed when perpendicular lines meet. Its measure is 90°.

A *straight* angle measures 180°.

An angle that measures 360° corresponds to a complete *revolution* of a ray about the vertex.

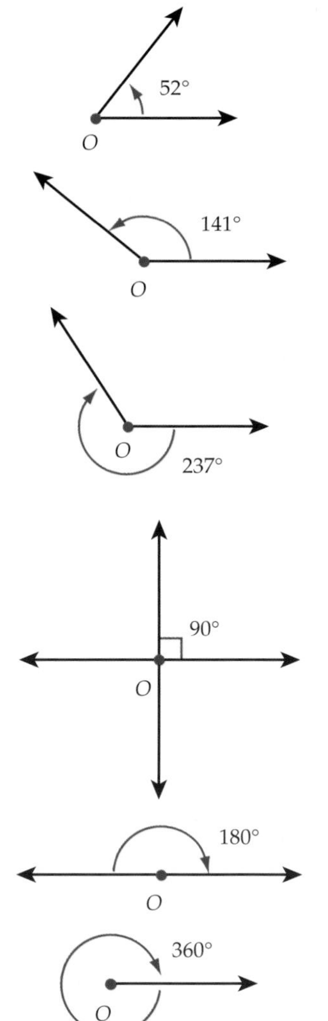

Lines and Angles

When several rays diverge from the same point, you have *angles at a point*, meaning all the angles have the same point as their *vertex*. The example below shows four rays diverging from point H.

You use three letters to indicate which angle you mean at point H. Simply saying, "angle H" is not sufficient because you cannot tell which of the four angles at H is being referred to.

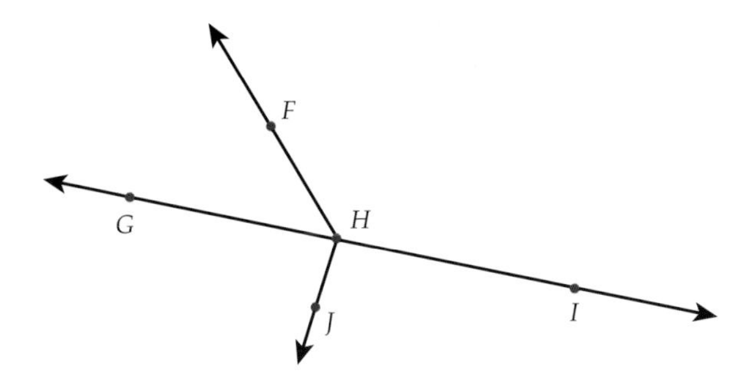

Angles at a point have measures that add up to 360°.

Two angles that have the same vertex and that have a ray in common are called *adjacent angles.* "Adjacent" things are things that are "next to" each other.

In the diagram above:

$\angle GHF$ and $\angle FHI$ are a pair of adjacent angles that together form a straight angle, $\angle GHI$.

$\angle FHG$ and $\angle GHJ$ are another pair of adjacent angles.

$\angle GHJ$ and $\angle JHI$ are a pair of adjacent angles that together form the straight angle $\angle GHI$.

Chapter 13

Defining Triangles

A *triangle* is a polygon with three sides formed by line segments.

Example

The triangle below is called ΔQRS.

The angles $\angle RQS$, $\angle QSR$, and $\angle SRQ$ are called the interior angles of ΔQRS.

Or, you can say $\angle Q$, $\angle R$, and $\angle S$ are interior angles of ΔQRS.

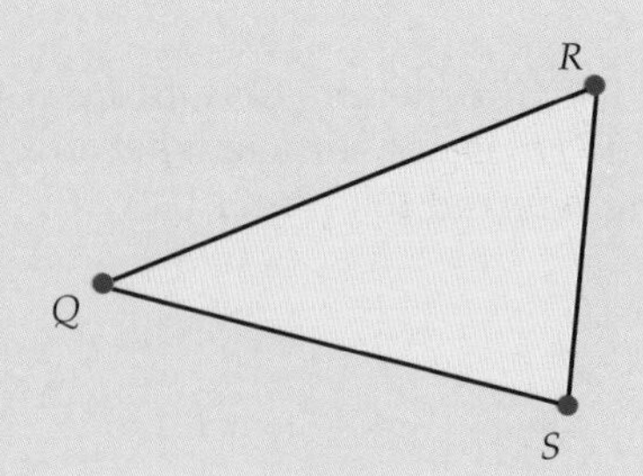

Congruent Sides

You can show that the sides of a triangle are congruent with congruence marks: | or ||.

Example

The sides $\overline{UV}$ and $\overline{UT}$ are the same length. This means $\overline{UV}$ and $\overline{UT}$ are congruent.

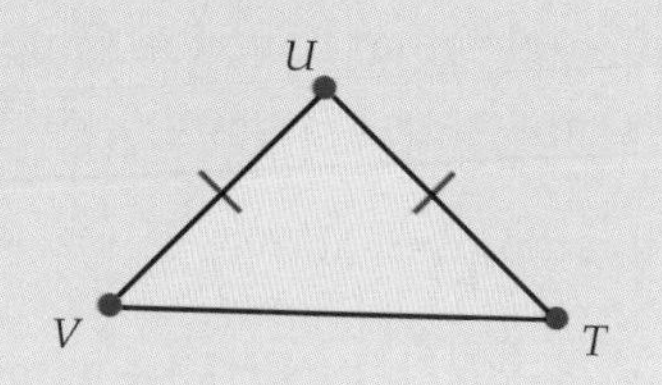

Sum of Interior Angles

The sum of the interior angles in any triangle is 180°.

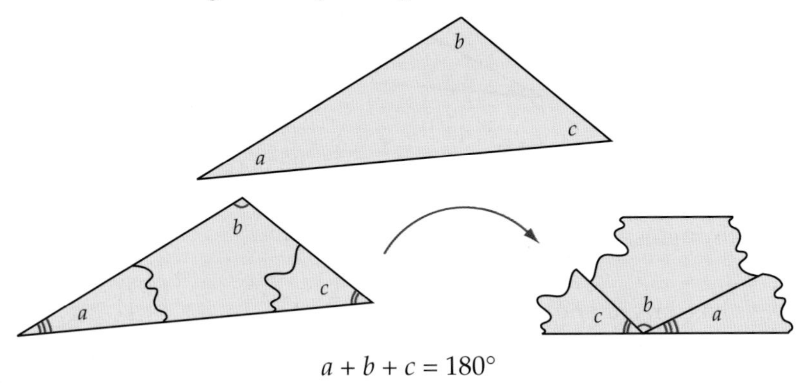

$a + b + c = 180°$

If you rearrange the triangles' vertices so they meet at a point, you can see the angles form a straight angle. The measure of the straight angle is 180°.

Types of Triangles

Triangles are a special type of polygon.

- Triangles have three sides and three angles.
- Triangles have the smallest number of sides and angles possible for a polygon.
- Triangles are always convex shapes.
- Every polygon that is not a triangle can be divided into triangular regions by sketching diagonals.

Example

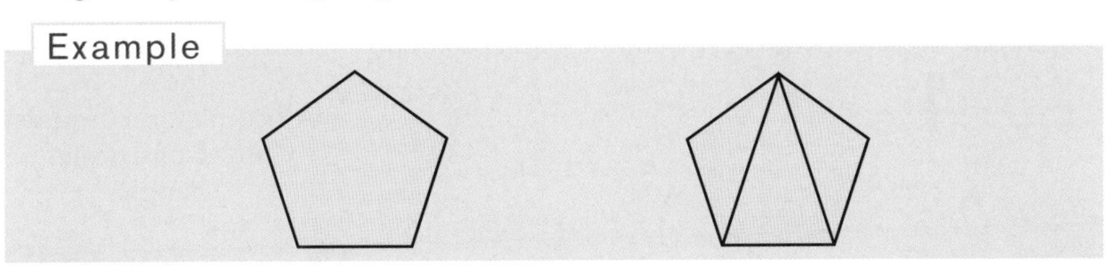

Triangles can be classified by the size of their angles.

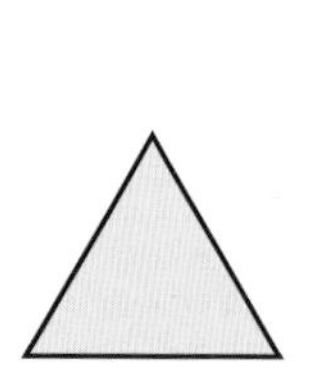
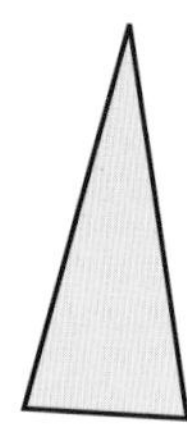
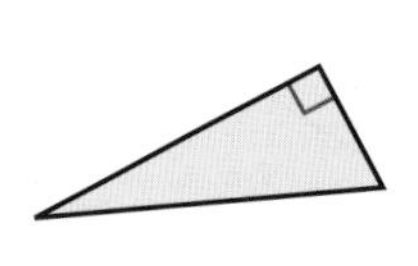
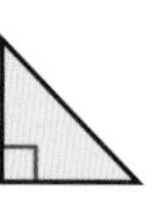
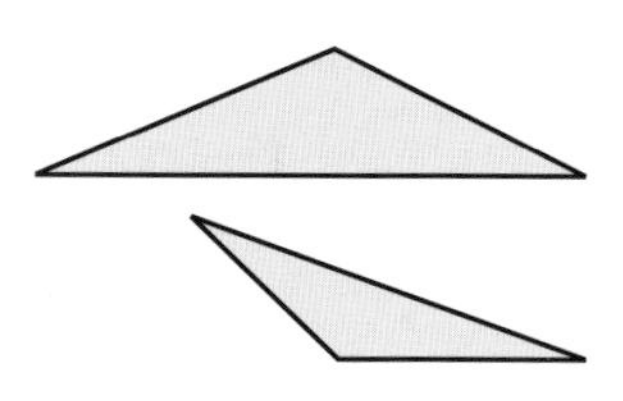

Acute triangles have three acute angles.

Right triangles have one right angle.

Obtuse triangles have one obtuse angle.

A *right triangle* has a right angle.
The longest side of a right triangle is the side opposite the right angle.

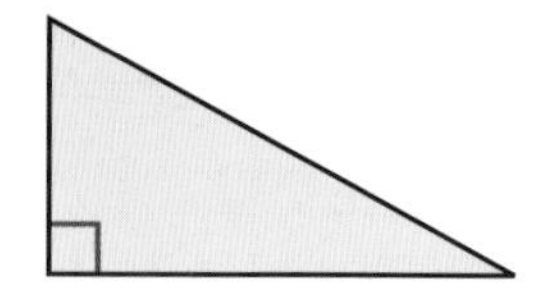

Triangles can be classified by the lengths of their sides.

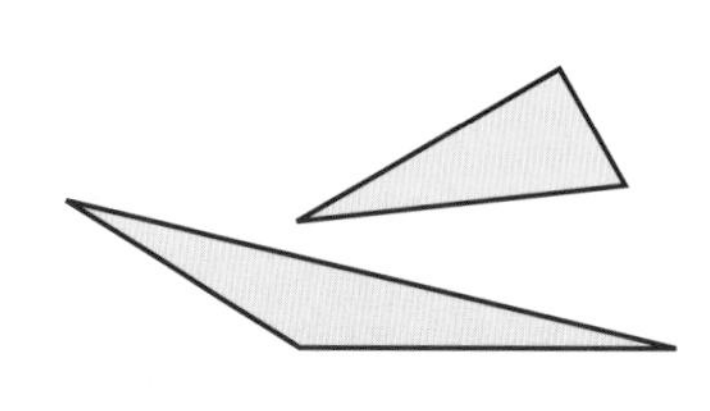
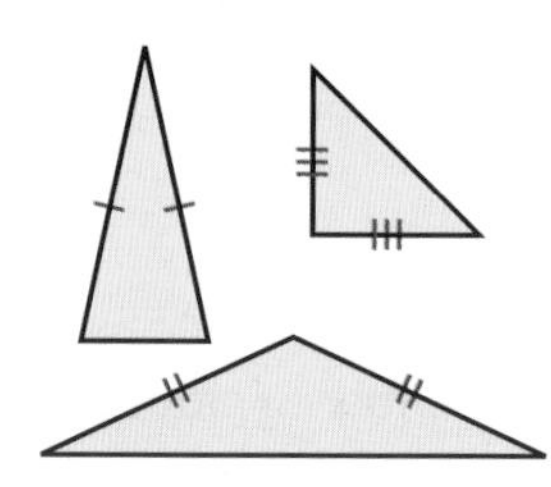
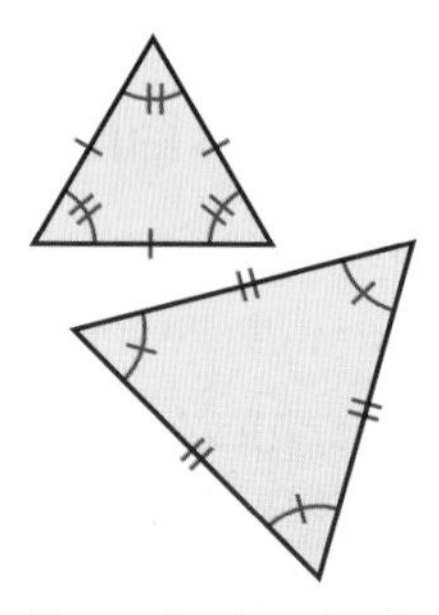

Scalene triangles have three sides of different lengths.

All three angles have different measures, too.

Isosceles triangles have two sides of the same length.

The angles opposite the equal sides are equal.

Equilateral triangles have three sides of the same length.

All angles are equal.

All equilateral triangles are also isosceles triangles.

Exterior Angles of a Triangle

For any triangle, the sum of the interior angles equals 180°. $\angle a + \angle b + \angle c = 180°$

An *exterior angle* makes a straight angle with its adjacent interior angle.

The sum of the exterior angle and its adjacent interior angle is 180°: $\angle e + \angle b = 180°$.

A set of exterior angles for a triangle consists of one exterior angle at each vertex.

The sum of the exterior angles in any triangle is 360°. (Rotating through the three exterior angles makes one complete revolution.)

This can be written as $\angle d + \angle e + \angle f = 360°$.

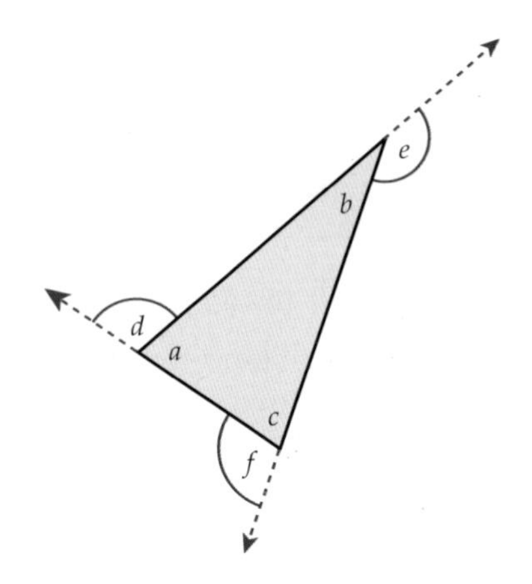

Calculating Angles of a Triangle

The sum of the measures of the three interior angles of a triangle is 180°. This will help you calculate unknown angles.

Example

In triangle STU, the measure of $\angle UTS$ is unknown. Let y stand for the measure of the unknown angle.

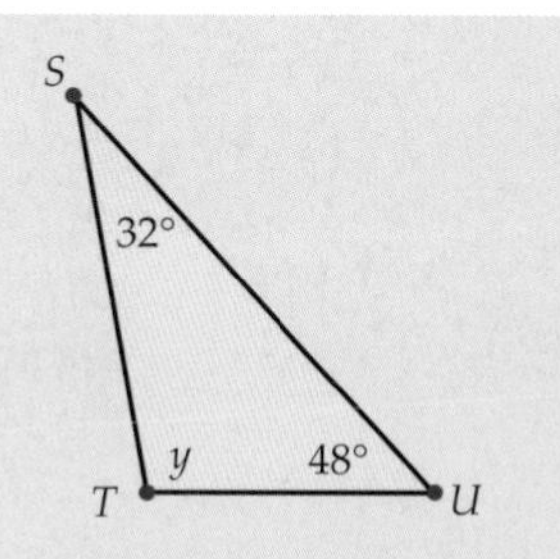

The sum of the measures of the interior angles of a triangle is 180°, so:

$$y + 48° + 32° = 180°$$

Solving this equation gives $y = 100°$. The measure of $\angle UTS$ is 100°.

In some cases, you may not know the measures of two angles.

Example

In triangle BCD, the measures of $\angle CBD$ and $\angle CDB$ are unknown.

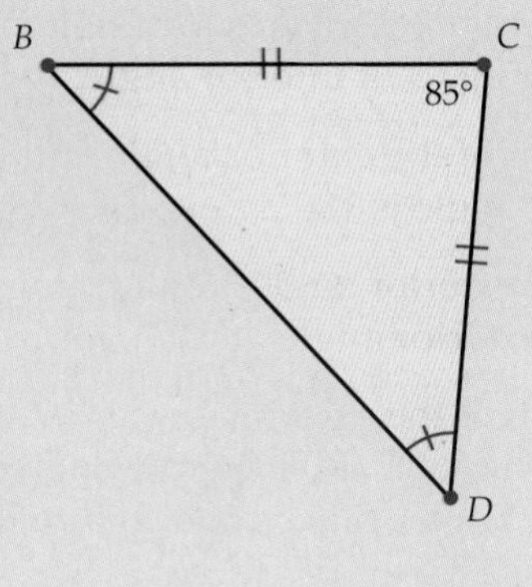

$\angle CBD$ and $\angle CDB$ are marked with the symbol for congruence, so $\angle CBD$ and $\angle CDB$ are congruent.

This means $\angle CBD = \angle CDB$.

Let x stand for the measure of these two unknown interior angles. The sum of the measures of the interior angles of the triangle is 180°, so:

$x + x + 85° = 180°$
$2x + 85° = 180°$
$2x = 95°$

Solving this equation gives $x = 47.5°$.
The measure of $\angle CBD$ is 47.5° and the measure of $\angle CDB$ is 47.5°.

Polygons

A *polygon* is a two-dimensional shape, or plane figure, that has these properties:

- All of the sides are line segments.
- The sides meet only at end points; they do not cross.

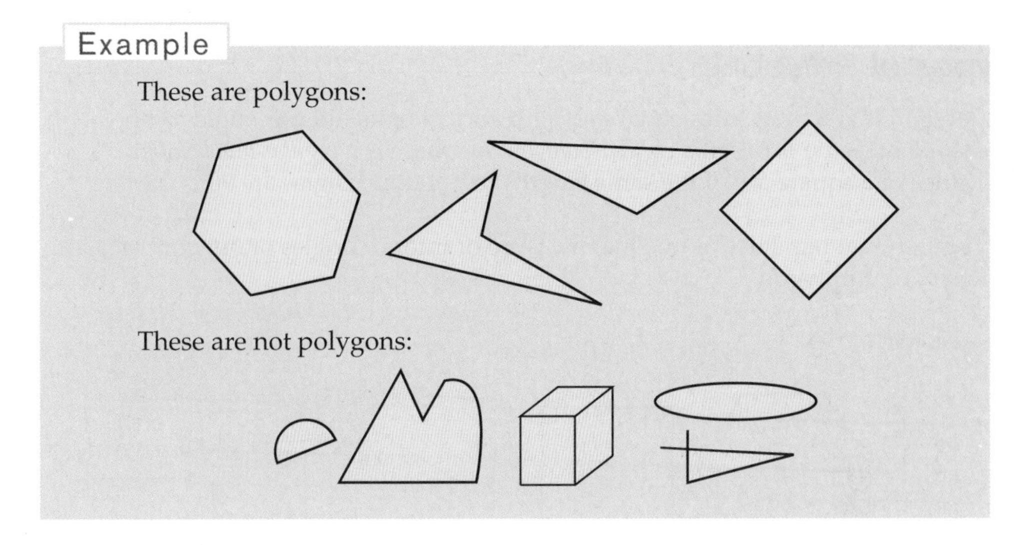

The line segments are the sides of the polygon. The point where two sides meet is called a *vertex*. The number of sides is always the same as the number of vertices.

Polygons with three sides are called *triangles;* polygons with four sides are called *quadrilaterals*. Some quadrilaterals have special names. Rectangles, squares, parallelograms, and trapezoids are examples of quadrilaterals. Polygons with five sides are *pentagons*. *Hexagons* are polygons with six sides. *Heptagons* are polygons with seven sides. *Octagons* are polygons with eight sides. *Decagons* are polygons with ten sides.

Sum of Interior Angles of a Polygon

The sum of the interior angles of a polygon depends on the number of sides of the polygon. It is the same for every polygon with that same number of sides. All triangles have an interior angle sum of 180°. All quadrilaterals have an interior angle sum of 360°. All pentagons have an interior angle sum of 540°.

Types of Polygons

A polygon that has all sides equal in length and all of its interior angles with the same measure is called a *regular polygon*. A square is a *regular quadrilateral*. All sides of a square are of the same length, and all four angles are 90°.

A polygon that has sides of unequal measure or interior angles of unequal measure is not a regular polygon.

Example

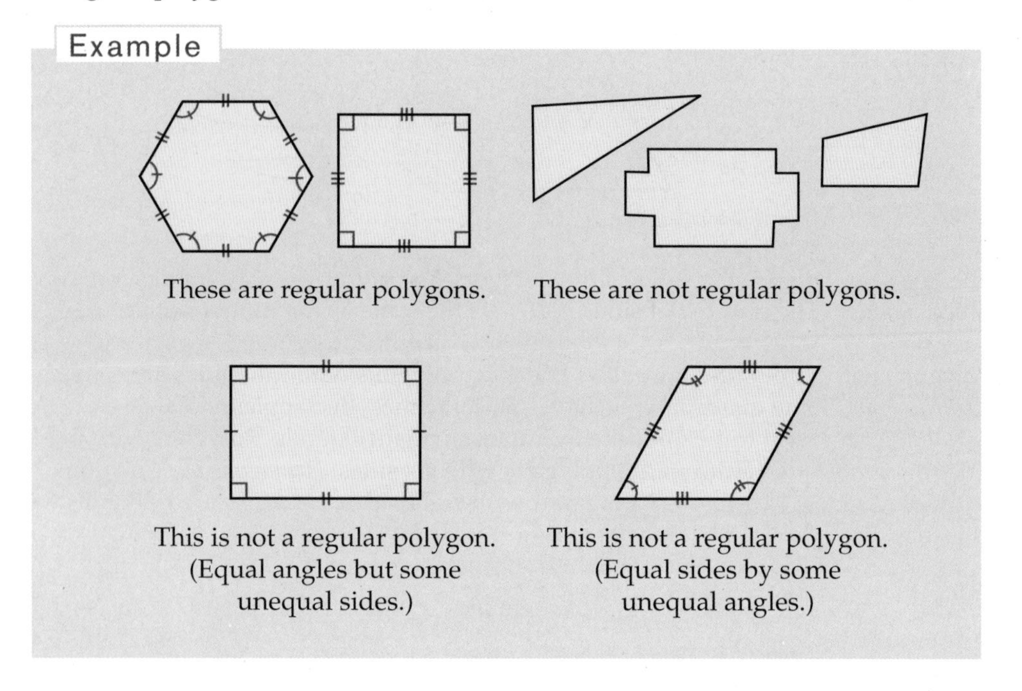

These are regular polygons. These are not regular polygons.

This is not a regular polygon. (Equal angles but some unequal sides.)

This is not a regular polygon. (Equal sides by some unequal angles.)

Polygons: Concave and Convex

A *concave polygon* is a polygon with at least one interior angle greater than 180°. It has a pushed-in, or caved-in, appearance.

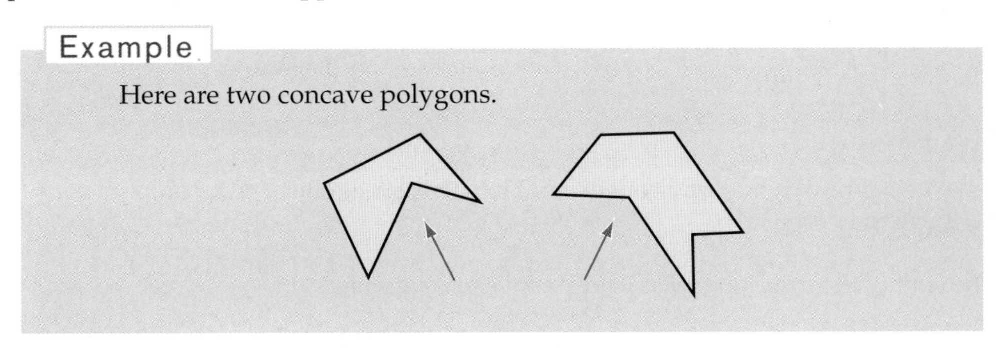

The pushed-in appearance is because one of the interior angles is a reflex angle. A concave polygon has at least one reflex angle.

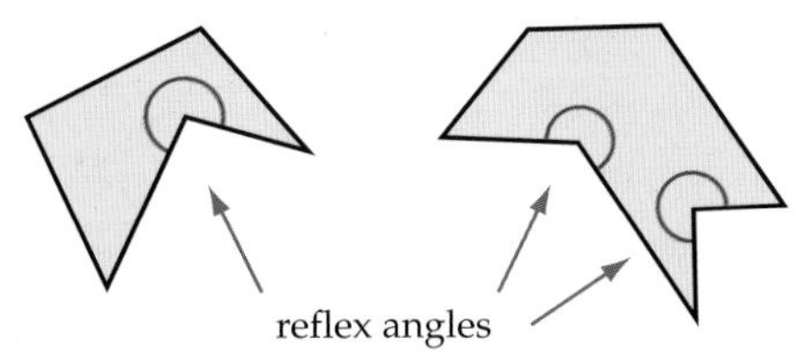

A polygon that is not concave is a *convex polygon*. In a convex polygon, no interior angles are reflex angles. Exterior angles can be drawn from all of the vertices.

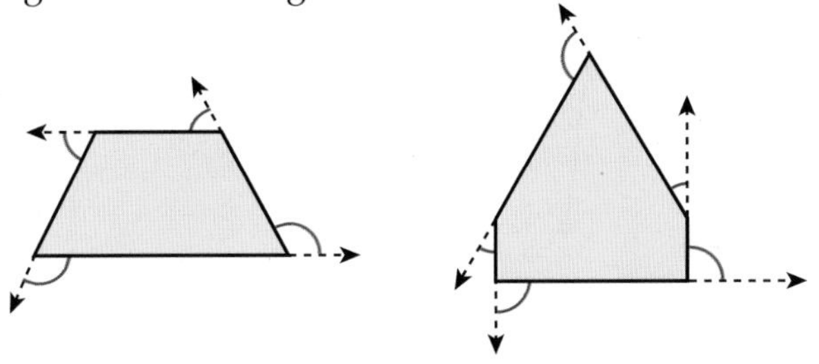

Exterior angles are the angles formed by extending the sides of a polygon into the region outside the polygon.

Chapter 13

Exterior Angles of a Convex Polygon

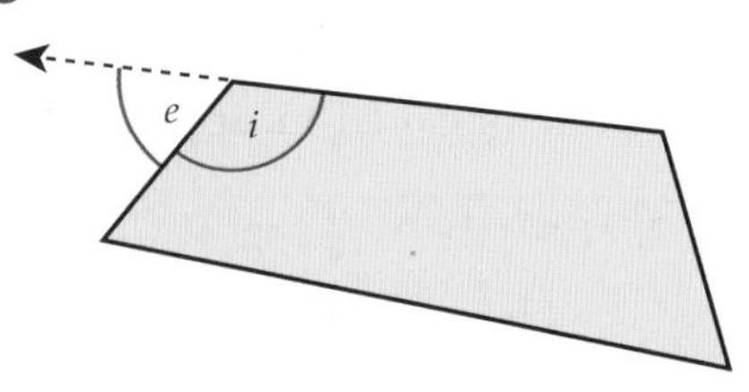

An *exterior angle* makes a straight angle with its adjacent interior angle.

The sum of the exterior angle and the interior angle is 180°.

$$\angle e + \angle i = 180°$$

A set of exterior angles for a convex polygon consists of one exterior angle at each vertex of the polygon.

The sum of the exterior angles in any convex polygon is 360°.

If you shrink a quadrilateral (or any polygon) down to just one point, you can see the angles sum to 360°. This is one way of explaining why the sum of the exterior angles of a quadrilateral is 360°.

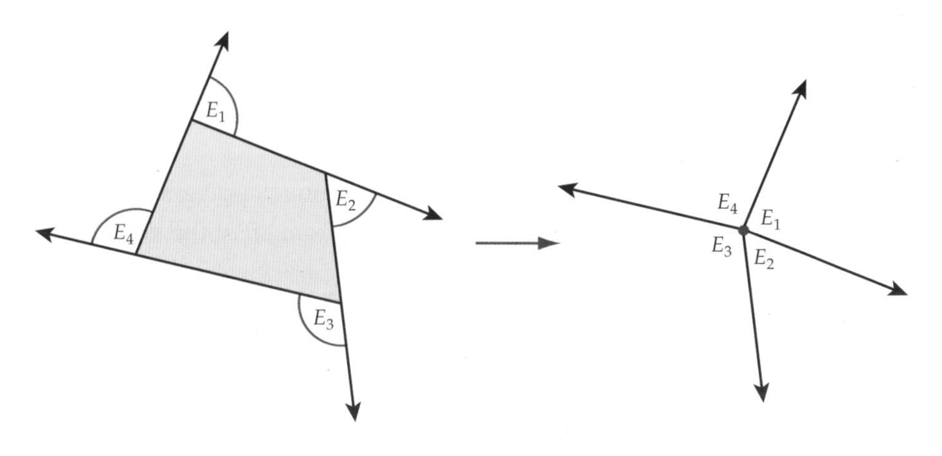

$$\angle E_1 + \angle E_2 + \angle E_3 + \angle E_4 = 360°$$

Quadrilaterals

Quadrilaterals are polygons with four sides.

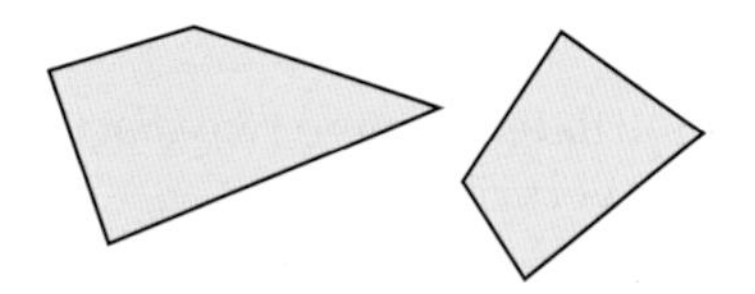

Types of Quadrilaterals

Parallelogram

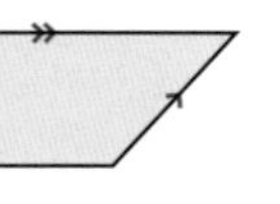

- 2 pairs of parallel sides
- Opposite sides of equal length

Rhombus

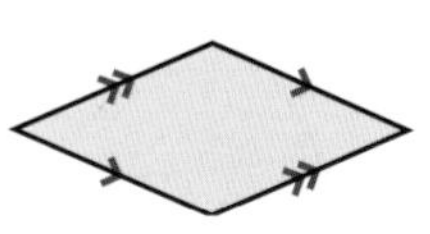

- 2 pairs of parallel sides
- 4 equal sides

Rectangle

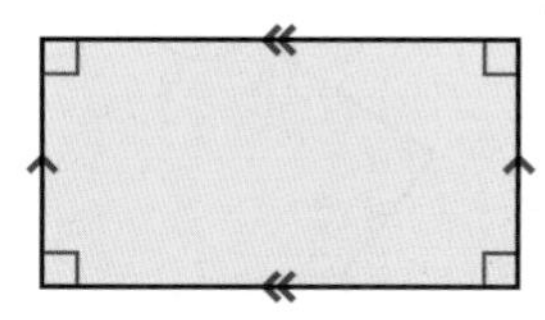

- 2 pairs of parallel sides
- 4 right angles
- Opposite sides of equal length

Square

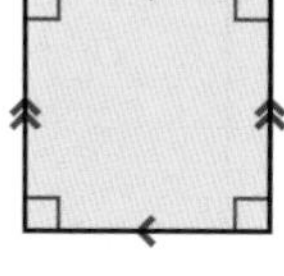

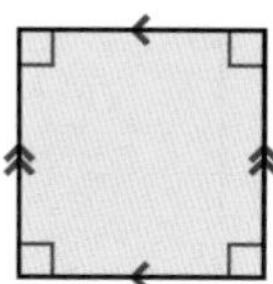

- 2 pairs of parallel sides
- 4 right angles
- 4 equal sides

Trapezoid

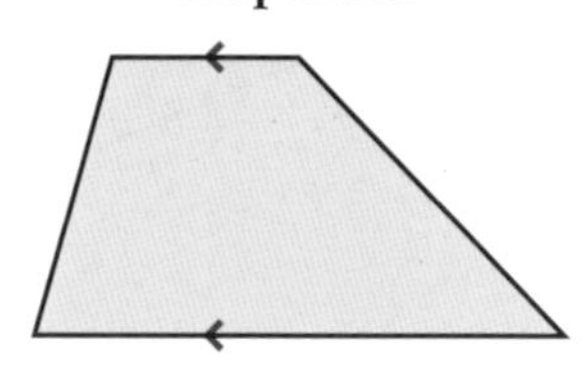

- 1 pair of parallel sides

Isosceles Trapezoid

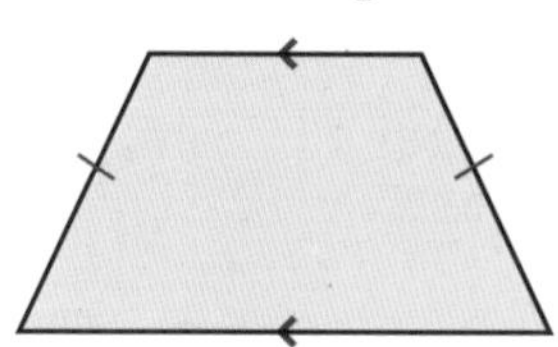

- 1 pair of parallel sides
- The other 2 sides of equal length

Kite

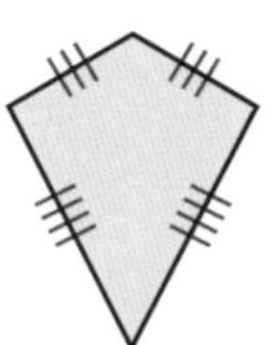

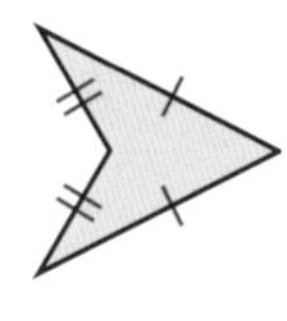

- 2 pairs of equal adjacent sides

Chapter 13

Quadrilaterals: Concave and Convex

All quadrilaterals have two *diagonals*, lines that can be drawn joining opposite vertices.

In a *convex* quadrilateral, both diagonals lie inside the quadrilateral, and each interior angle is less than 180°. The number of obtuse angles can be zero, one, two, or three.

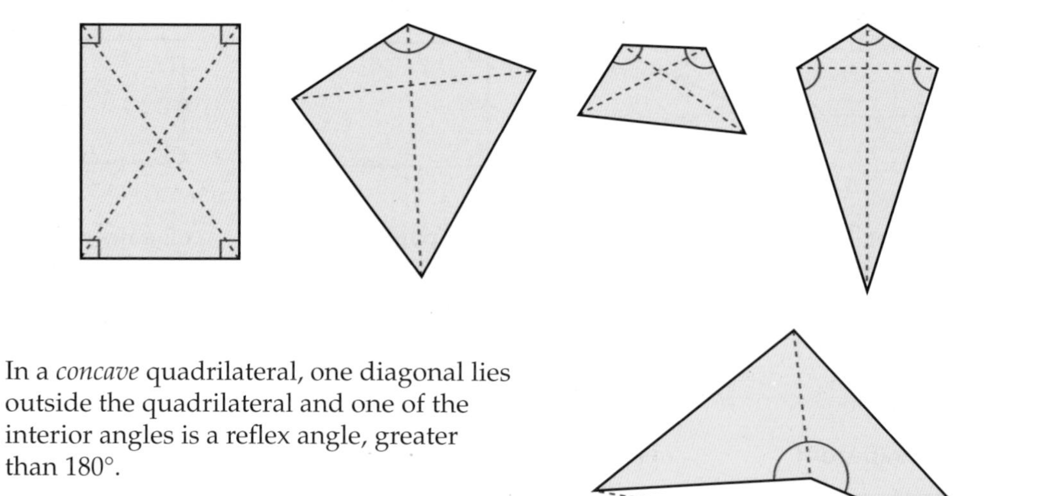

In a *concave* quadrilateral, one diagonal lies outside the quadrilateral and one of the interior angles is a reflex angle, greater than 180°.

Sum of Interior Angles of a Quadrilateral

An internal diagonal divides a quadrilateral into two triangles. This fact can be used to prove that the sum of the interior angle measures of a quadrilateral is 360°.

The following proof assumes that you already know that the sum of the angle measures for a triangle is 180°.

Angle sum

$= (\angle A + \angle D) + \angle B + (\angle C + \angle F) + \angle E$

$= (\angle A + \angle B + \angle C) + (\angle D + \angle E + \angle F)$

$= 180° + 180°$

$= 360°$

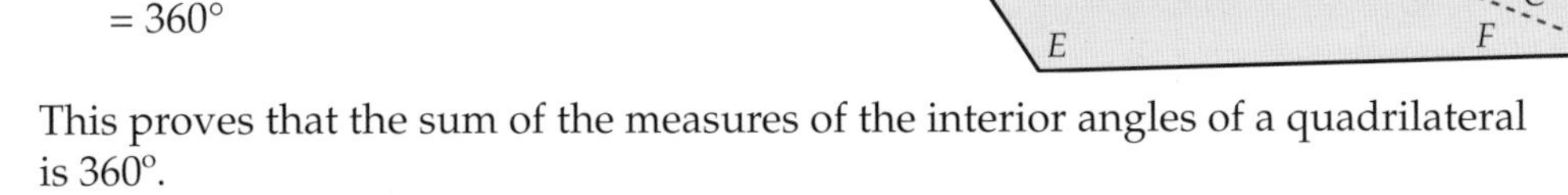

This proves that the sum of the measures of the interior angles of a quadrilateral is 360°.

Testing Conjectures about Geometric Figures

A conjecture is a statement which you need to check to see whether or not it is true. How can you what you have learned about different properities to make and test conjectures about geometric figures?

Here is a conjecture:

"An equilateral triangle can also be a right triangle."

Think about the properties of an equilateral triangle.

- An equilateral triangle has 3 equal sides and 3 equal angles.
- Each angle of an equilateral triangle is 60°.
- The sum of the interior angles of a triangle equals 180°.
- A right triangle has one angle equal to 90°.

Therefore, the conjecture is false.

An equilateral triangle can never be a right triangle. You know this because the angles of an equilateral triangle all have the same measure of 60°. But a right triangle must have one angle that measures 90°.

GEOMETRIC MEASURE

CHAPTER 14

Length: A One-Dimensional Measure

The general term *length* is used for any measurement that can be made directly with a ruler or tape measure (or indirectly with something like a car odometer).

Length is a measure of distance, a one-dimensional measure. Later in this chapter is a discussion of area, a two-dimensional measure, and volume, a three-dimensional measure.

To illustrate the idea of measuring length, imagine a brick. A standard brick is about 8 inches long, 4 inches wide, and 2.5 inches high.

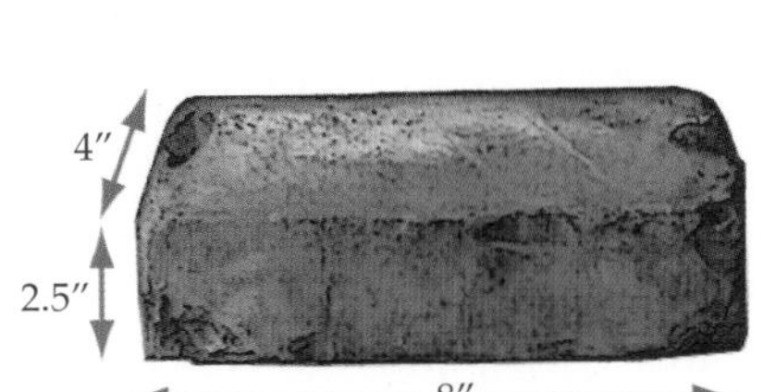

The 8-inch measurement is called the *length,* the 4-inch measurement is called the *width,* and the 2.5-inch measurement is called the *height.* Each of these are examples of one-dimensional measurements of length.

(Notice that the term *length* is used in two ways, one refers specifically to the long dimension of the brick as opposed to its width and height, and the other refers generally to any one-dimensional measure.)

All measurements require a *unit*. Length is typically measured either in English units such as inches, feet, or miles, or in metric units such as centimeters, meters, or kilometers. But the units could also be *nonstandard* units you make up (such as the number of paper clips laid end to end, or the number of your hand widths).

The unit that you choose should be appropriate for the particular item being measured. Sometimes, more than one unit is appropriate. Heights of people, for example, can be measured in inches or in feet and inches.

All measurements are approximations; however, the more detailed the scale you use, the more *accurately* you can measure. Using the scale below, the most accurate measurement you can make for the shaded line is $4\frac{1}{8}$ inches.

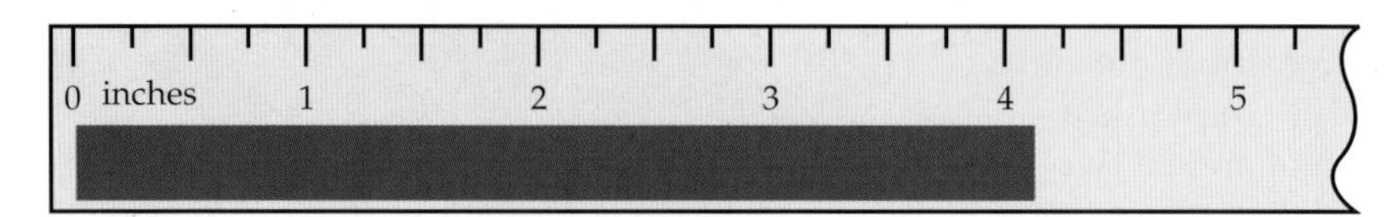

Three Types of Length Measurements

There are three main ways you can describe the size of this rectangle.

Distance along: If you lay a ruler along its sides, you find its length to be 4 cm and its width to be 3 cm. Here, length is shown as a measure of distance *along* an edge of an object.

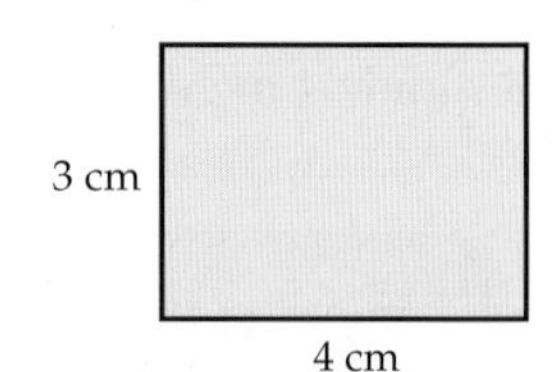

Distance across: If you lay a ruler from one corner to the opposite corner along the diagonal, you find this distance to be 5 cm. Here, length is shown as a measure of distance *across* an object.

3 cm

5 cm

4 cm

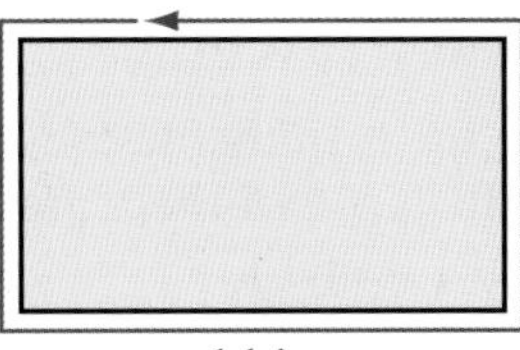

Distance around: If you wrap a string all the way around the rectangle, and then lay the string out and measure it, you find this distance to be 14 cm. Here, length is shown as a measure of distance *around* an object.

The distance around a figure is usually called *perimeter,* but for a circle it is called *circumference.*

Distances along, across, or around are all *one-dimensional* measures of length. It is helpful to think of a string when you think of one-dimensional measures. Though it is a three-dimensional object, you can think of string as having just one dimension—its length. A piece of string has the same length whether you lay it out straight, make it turn corners, or follow a curve.

Distances along, across, and around are also called *linear* measures, because it is possible to draw a straight line to represent them. (Whenever you use a string to measure something, it can be stretched out in a straight line.)

One-dimensional measures can be made on one-, two-, or three-dimensional figures. More examples appear later in this chapter. The linear measures have different names depending on their use: *length, width, height, radius, diameter, perimeter, circumference,* and *girth.*

Unit Conversion of Length

Measurement can be in the English system (inches, feet, miles) or metric system (centimeters, meters, kilometers).

You can convert a measurement to a different unit within the same system, or you can convert between different systems.

Here, you will focus on conversions within the same system of measurement.

English System

Length measurements can be converted in the English system from feet to inches using the fact that there are twelve inches (written 12 in or 12") in one foot (written 1 ft or 1').

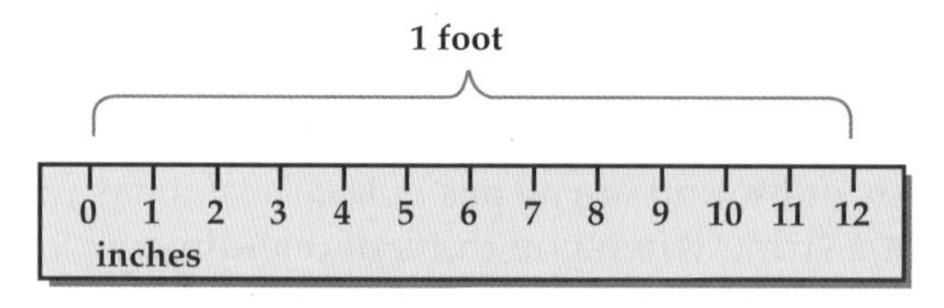

Here are some conversions between different English units for length measurements:

1 foot (ft) = 12 in   1 yard (yd) = 3 ft   1 mile (mi) = 1,760 yd = 5,280 ft

Example

A student is 5 feet tall. What is the student's height in inches?

Because there are 12 inches in one foot, the student's height is

$$5 \bullet 12 \text{ in} = 60 \text{ in}$$

Metric System

Length measurements can be easily converted within the metric system.

10 millimeters (mm) = 1 centimeter (cm)

100 centimeters (cm) = 1 meter (m)

1,000 meters (m) = 1 kilometer (km)

Distance Across

A *diagonal* of a rectangle is a line segment joining opposite corners.

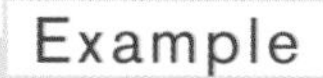

Example

If you use a ruler to measure the diagonal of this rectangle, you get a length of 6.7 centimeters, or 67 millimeters.

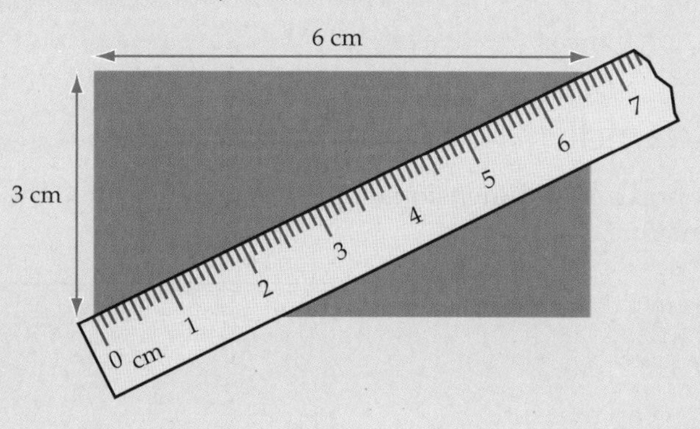

A line segment from one point on a circle to any other point, passing through the center, is a *diameter* of the circle.

Example

In this circle, line segment AB is a diameter that passes through the center, O.

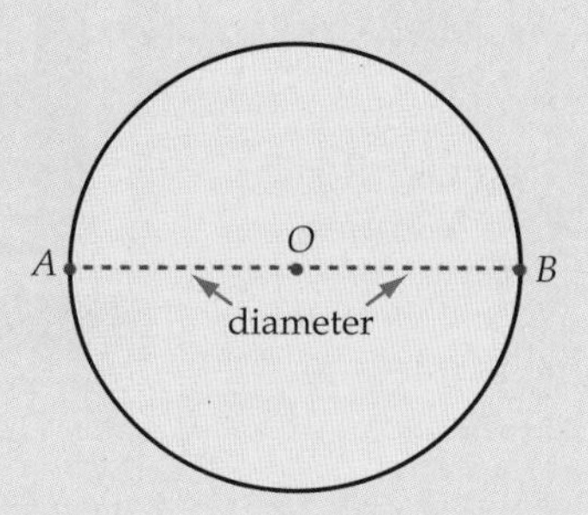

A line segment from the center to any point on a circle is a *radius* of a circle.

In this circle, line segment *OE* is a radius.

The diameter of a circle is equal to twice the radius. This means $d = 2 \bullet r$.

The diameter and the radius are both one-dimensional measures of a circle.

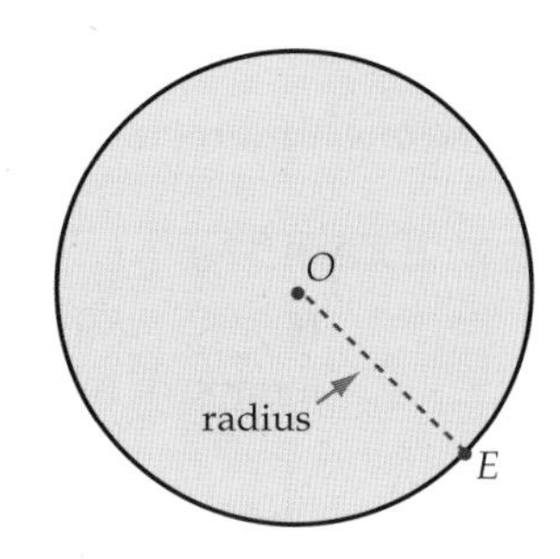

Distance Around

The distance around a two-dimensional figure is usually called the *perimeter*, a one-dimensional measure.

The distance around this rectangle is called its perimeter.

Using a ruler, you can measure the perimeter of a rectangle in inches or centimeters. Each side length is measured, and the measures are added.

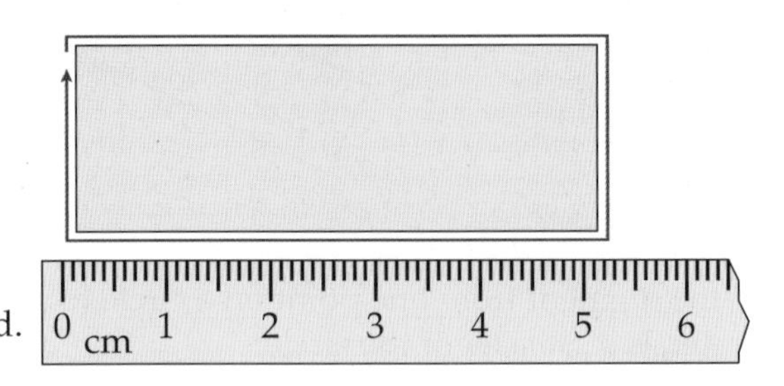

Example

Suppose the street around a college campus is called Perimeter Road.

The length of Perimeter Road could be measured in miles, meters, or feet.

Each of these measures is a one-dimensional measure of length.

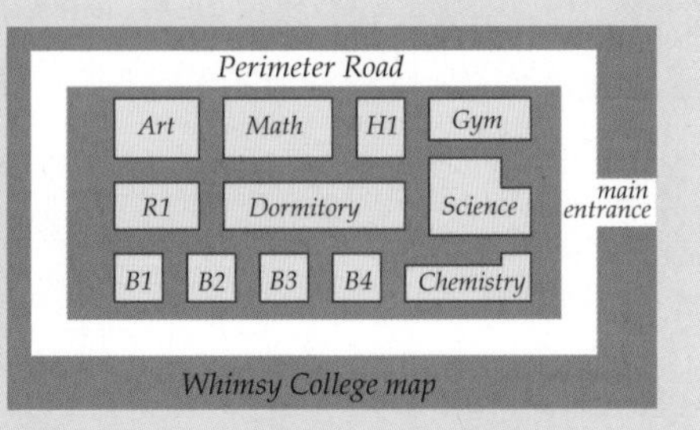

Example

The perimeter of the outside square shown here is $4 \bullet 5 = 20$ cm.

The perimeter of the shaded shape is found by adding side lengths.

The square and the shaded polygon have equal perimeters. Can you see why? Notice they have the same opposite corners at points R and Q.

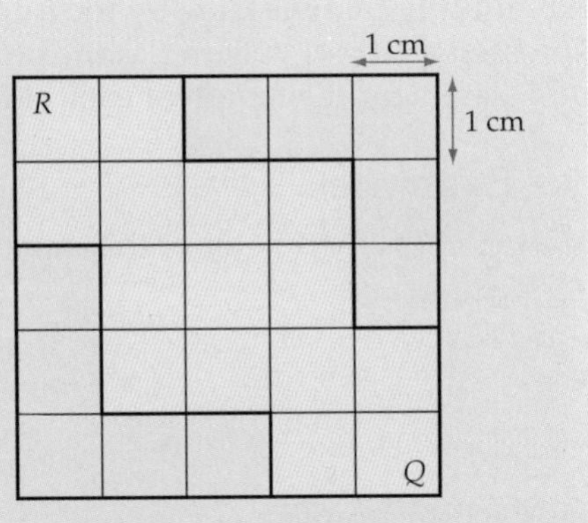

Starting from point R, the perimeter of the shaded shape is:
$2 + 1 + 2 + 2 + 1 + 2 + 2 + 1 + 2 + 2 + 1 + 2 = 20$ cm

You can calculate the perimeter of a polygon by adding the lengths of the sides.

- In some problems, you calculate the perimeter by adding given measures. Before adding, it is important to check that the units are the same for each side.
- For some problems, like the one below, you need to make direct measures of the sides (using an appropriate unit), and then add them.

Perimeter = sum of measured lengths = $a + b + c + d$

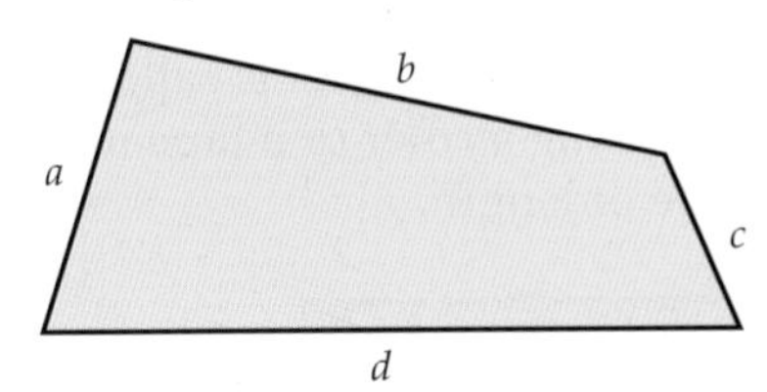

To determine the perimeter of a regular polygon, you need to measure only one side and multiply that measure by the number of sides. Using letters, this rule can be expressed as $P = nl$, where P stands for perimeter, n stands for the number of sides, and l stands for the length of each side.

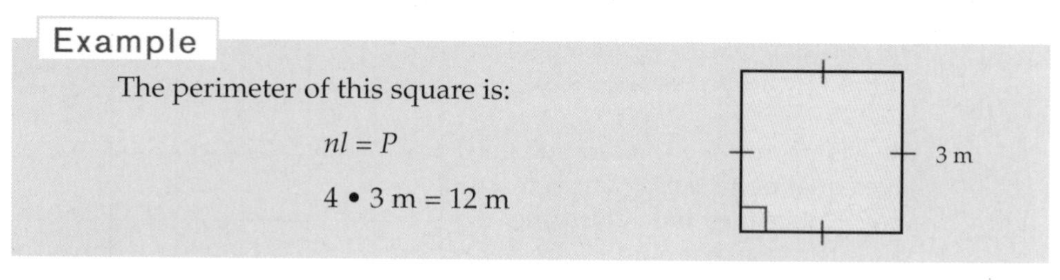

Example

The perimeter of this square is:

$$nl = P$$

$$4 \bullet 3\text{ m} = 12\text{ m}$$

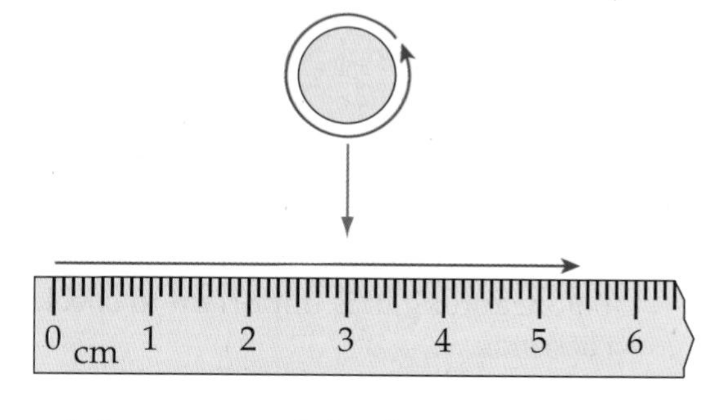

The distance around a circle is called the *circumference.* The circumference of a small circular object can be measured in inches or centimeters using a piece of string and a ruler.

First, you carefully lay the string around the object and pull it tight. Then you use a ruler to measure the length of the string that is needed to go around the object.

Circumference is a one-dimensional measure of length.

The distance around a three-dimensional object is called the *girth.* Girth can be measured directly using a ruler or a tape measure. It is a one-dimensional measure of length.

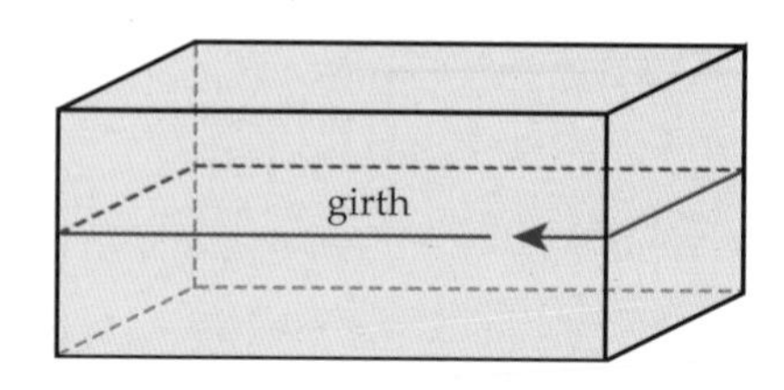

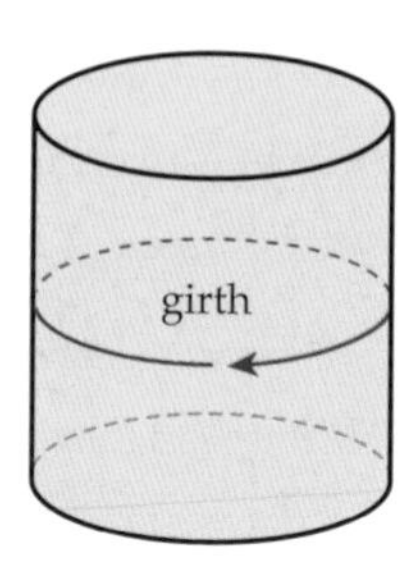

Circumference

The perimeter of a circle has a special name—the *circumference*.
The circumference of a circle can be measured directly. The example shows two ways.

Example

1. Wrap a tape measure around the circular object and read the measurement.

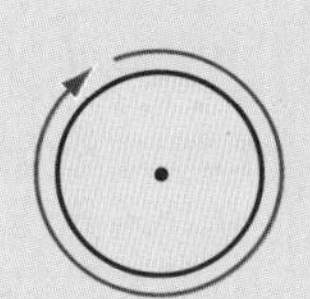

2. Roll the circular object for one complete turn along a flat surface and measure the distance traveled.

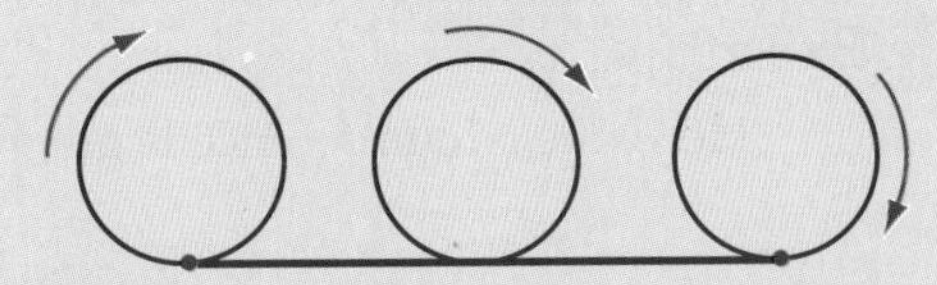

Another way to find the circumference of a circle is to use your knowledge of circles to calculate circle measurements.
If you measure the circumference of a circle and compare it to the length of the diameter of the same circle, you will see a constant relationship:

The circumference is just a little more than 3 diameters long.

The ratio of the circumference, C, to the diameter, d, is π. That means $\frac{C}{d} = \pi$.

The symbol π (spelled "pi" and pronounced "pie") represents an irrational number between 3.1 and 3.2. You cannot write down its exact value, so you call it π.
For many calculations, it is acceptable to use the approximation $\pi \approx 3.14$.

You rely on the ratio $\frac{C}{d} = \pi$ when you want to calculate circle measurements.

If you know the radius, the diameter, or the circumference of a circle, then you can calculate the other two.

Example

Here, the radius is given:

Diameter, $d = 2r = 2 \bullet 9 = 18$ cm

Circumference, $C = \pi d$
$C = \pi 18 \approx 56.5$ cm

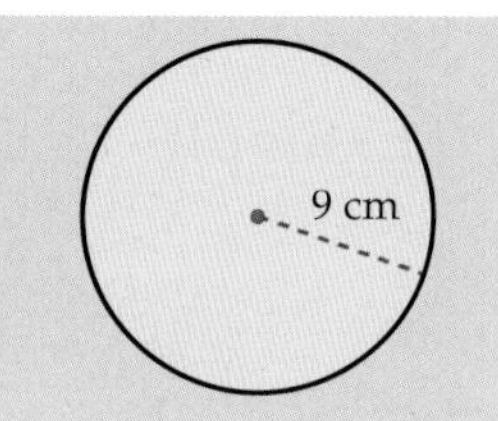

Example

Here, the circumference is given.

Diameter, $d = \frac{C}{\pi} \approx \frac{5}{3.14} \approx 1.59$ in

Radius, $r = \frac{C}{2\pi} \approx \frac{5}{2 \bullet 3.14} \approx 0.796$ in

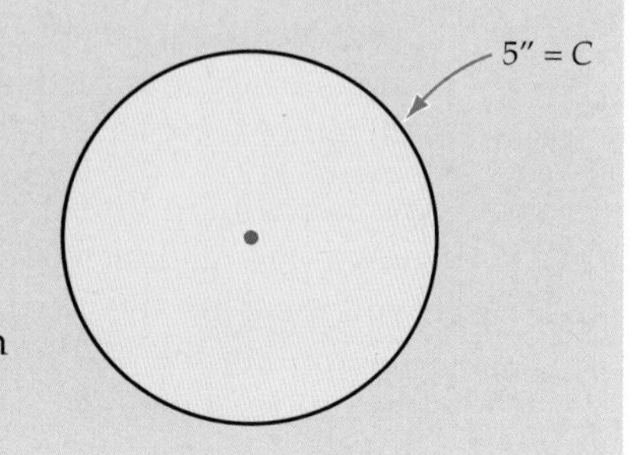

Area: A Two-Dimensional Measure

Area is a measure of the size of two-dimensional figures or of the size of the surfaces of three-dimensional solids.

Area is a *two-dimensional* measure. It may be helpful to think of the two-dimensional nature of area in terms of *cloth.* You can think of cloth as having just two dimensions—length and width. (Its third dimension, thickness, is usually very small compared with its length and width.) An important way to measure the *size* of a piece of cloth is its *area.* In fact, material is often sold by the square yard, a measure of its area.

Area is measured in *two-dimensional* units. The unit can be any size, as long as it is two-dimensional. Area is the number of *square* units in a figure.

Here is a rectangle covered with 50 one-centimeter square tiles.

1 cm

1 cm

1	2	3	4	5	6	7	8	9	10
11	12	13	14	15	16	17	18	19	20
21	22	23	24	25	26	27	28	29	30
31	32	33	34	35	36	37	38	39	40
41	42	43	44	45	46	47	48	49	50

The area of the rectangle is 50 square centimeters. You are using a standard unit of measurement, centimeters. The area is written as 50 cm^2. This measure has the same meaning for everyone, with or without the tiles as an illustration.

The following rectangle is covered with square tiles of an unknown side length. The tiles are all the same size.

1	2	3	4	5	6
7	8	9	10	11	12
13	14	15	16	17	18

The area is 18 tiles, where a "tile" is a nonstandard unit of area.

In this case, you are measuring area with a unit that is not defined as a standard unit of measurement. You know the area is 18 tiles, but you do not know the actual measurement of the tiles.

Chapter 14

Area of a Rectangle

You do not want to have to use tiles to cover a plot of farmland, the ceiling of a room, or the area of a window opening. Besides, most rectangles cannot be covered exactly with one-inch or one-centimeter tiles.

To find the area of a rectangle (a two-dimensional measure) without counting squares, you multiply its width by its height.

This is represented by the formula Area = length • width, or $A = lw$.

Find the area of a 10 cm × 6 cm rectangle.

$A = lw$

$A = 10 \bullet 6 = 60 \text{ cm}^2$

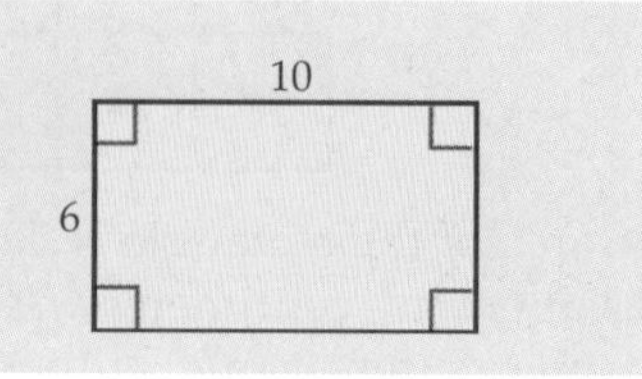

Another name for the length of a rectangle is *base*, and another name for the width is *height*. The height of a figure is defined as the perpendicular distance from the top of the figure to the base.

Area of a Triangle

Any rectangle can be cut up and rearranged into two congruent right triangles. This means that the area of each of the right triangles must be *half* the product of the rectangle's base times the height.

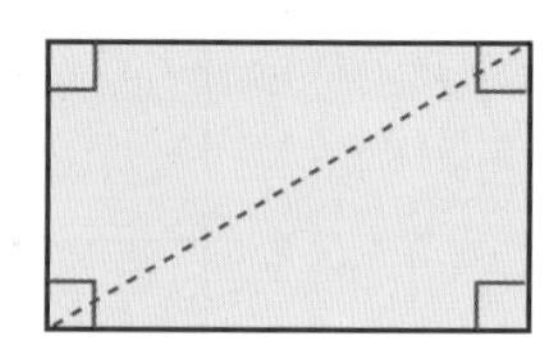

In fact, the area of *any* triangle is half the product of the base times the height, which is represented by the formula for the area of a triangle:

$$A = \frac{1}{2}bh$$

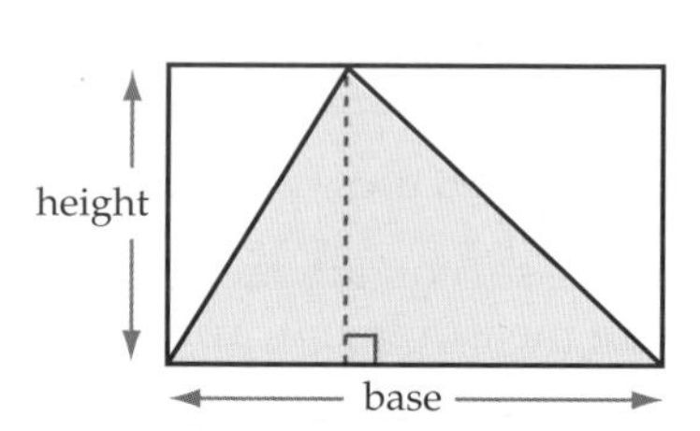

Remember, the product of two linear measures gives a two-dimensional measure, area.

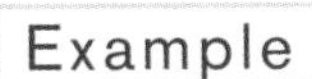

For this right triangle:

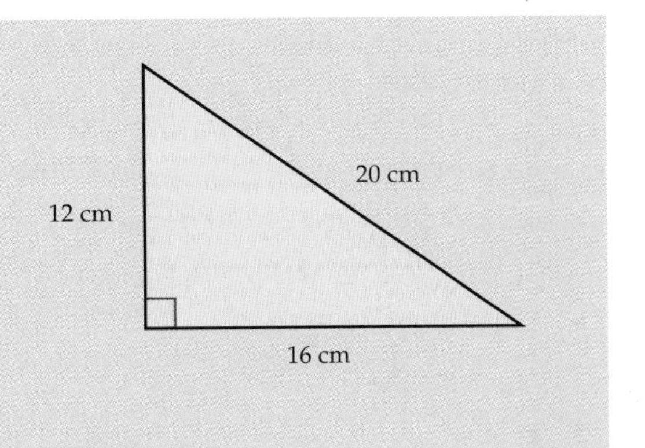

$$A = \frac{1}{2}bh$$

$b = 16 \qquad h = 12$

$$A = \frac{1}{2} \bullet 16 \bullet 12$$

$A = 96\text{ cm}^2$

In the same triangle, if the 20 cm side is chosen as the base, then the height is different. This height can be calculated using the known value of the area.

Example

For the same right triangle:

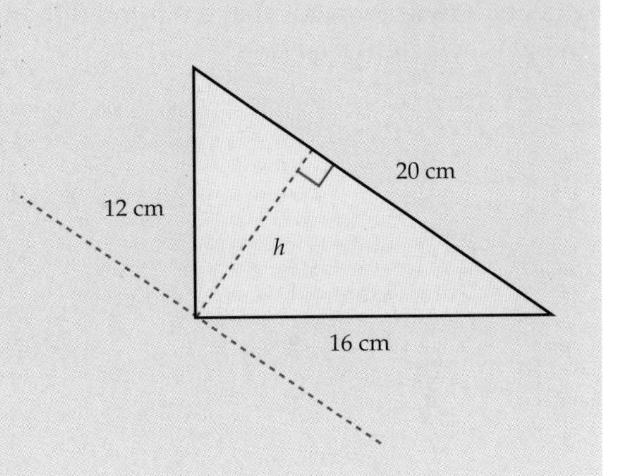

$$A = \frac{1}{2}bh$$

$$96 = \frac{1}{2} \bullet 20 \bullet h$$

$96 = 10 \bullet h$

$96\text{ cm}^2 \div 10\text{ cm} = h$

$9.6\text{ cm} = h$

Area of a Polygon

Knowing the formulas for the area of a triangle and the area of a rectangle allows you to determine the area of all other polygons, regular and irregular. Do this by dividing the polygon into combinations of right triangles and rectangles, and then adding the areas of the component shapes.

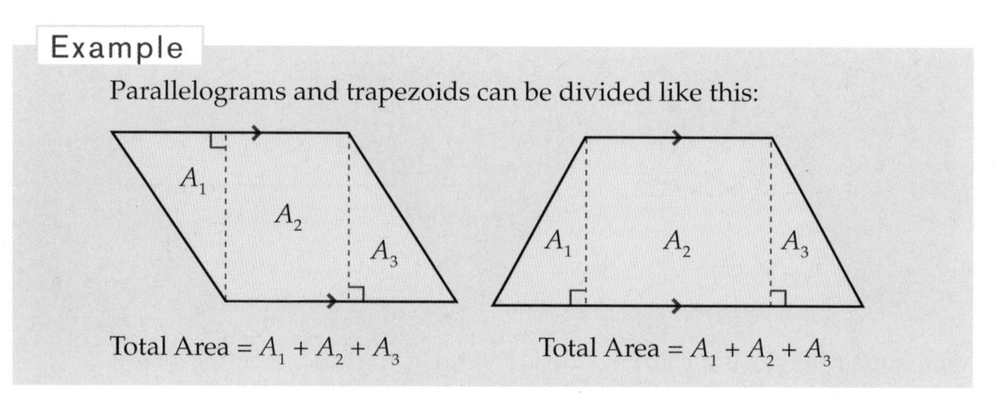

For all other polygons, a good method is to first construct any one diagonal of the polygon. Then join each of the other vertices to the diagonal with perpendicular line segments. Any trapezoids that are formed in this process can then be subdivided into rectangles and right triangles.

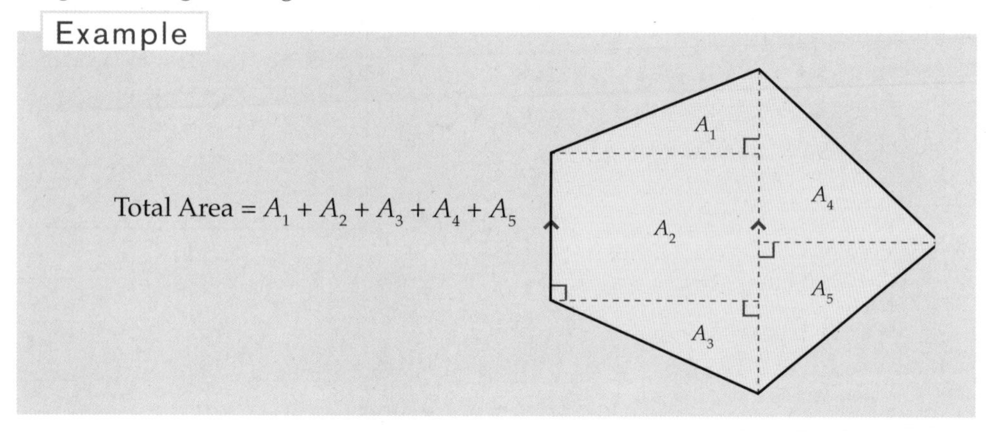

Area of a Circle

In any circle, the ratio of the area, A, to the square of the radius, r^2, is π: $\frac{A}{r^2} = \pi$.

You can find the area, A, by multiplying the square of the radius, r^2, by π: $A = \pi r^2$.

Here, area involves squaring a linear measure, the radius.

Example

Suppose the radius of a circle is exactly 8 meters.

$A = \pi r^2$

$A = \pi \bullet 8^2$

$A = \pi \bullet 8 \bullet 8$

$A = 64\pi$ m^2 (exactly)

$A \approx 64 \bullet 3.14 \approx 201$ m^2 (approximately)

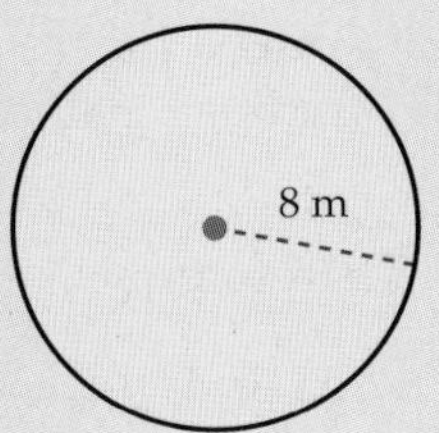

Area of a Composite Figure

A composite figure is a two-dimensional shape that combines two or more simpler shapes. You can calculate the area by dividing the composite figure into smaller shapes. These individual areas are calculated and added to give the total area. Look for simple shapes like triangles, rectangles, squares, circles, and semicircles.

Example

This composite figure is made up of a semicircle, a square, and a triangle.

Total area = $A_1 + A_2 + A_3$

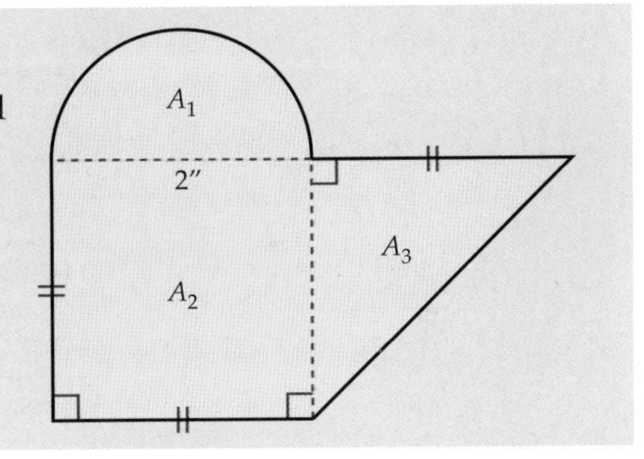

Calculate and add the areas of the smaller shapes to get the total area.

Example

Figure A_1 is a semicircle.

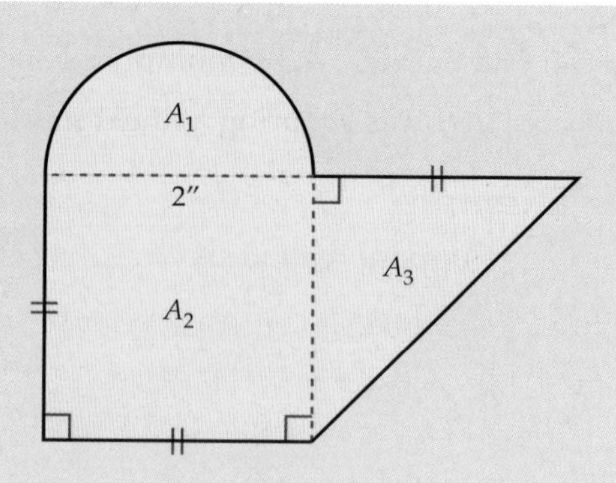

$$\text{Area} = \frac{1}{2}\pi r^2$$

Diameter = 2 inches

Radius = 1 inch
(radius = half of diameter)

Therefore,

$$\text{Area of } A_1 = \frac{1}{2}\pi r^2 = \frac{1}{2} \bullet \pi \bullet 1 \approx \frac{1}{2} \bullet 3.14 \approx 1.57 \text{ in}^2$$

Figure A_2 is a square.

Area = s^2

Area of $A_2 = 2^2 = 4$ in^2

Figure A_3 is a triangle.

$\text{Area} = \frac{1}{2}bh$ You know that $b = 2$ and $h = 2$.

$$\text{Area of } A_3 = \frac{1}{2} \bullet 2 \bullet 2 = 2 \text{ in}^2$$

Add all the areas to get the total area:

$$A_1 + A_2 + A_3 \approx 1.57 + 4 + 2 = 7.57 \text{ in}^2$$

Surface Area

The *surface area* is the sum of the areas of all the exterior surfaces of a solid. Surface area is a two-dimensional measure of a three-dimensional figure.

Surface area is measured in two-dimensional units such as square millimeters (mm^2), square centimeters (cm^2), square inches (in^2), square feet (ft^2), or square kilometers (km^2).

Surface Area of a Rectangular Prism

A rectangular prism is a prism in which all faces are rectangles. A rectangular prism is like a box. Consider this box. There are many ways to cut along the edges to flatten the box.

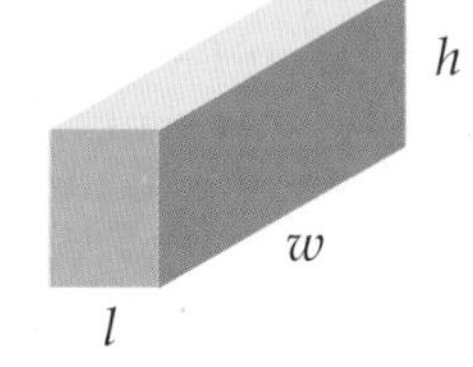

Compare these two different ways to flatten this box.

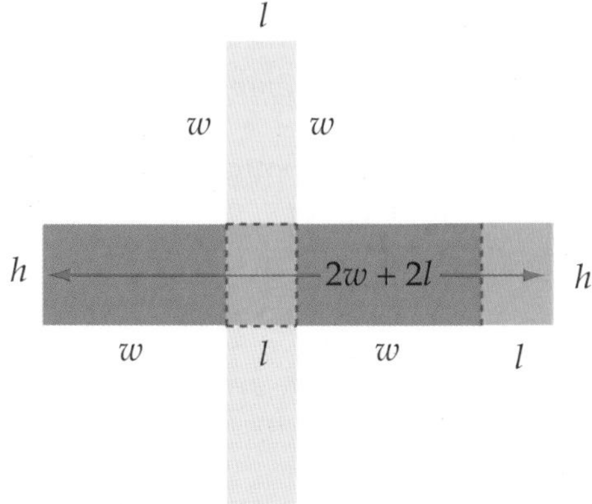

w
h
2h + 2l
l
h

Total surface area

$= h(2w + 2l) + 2lw$

$= 2hw + 2hl + 2lw$

$= 2(hw + hl + lw)$

Total surface area

$=(2h + 2l)w + 2(hl)$

$= 2hw + 2lw + 2hl$

$= 2(hw + hl + lw)$

Both ways of flattening this box show that the surface area of the box is $2(hw + hl + lw)$. The surface area is the total area of the six rectangles that form the six faces of the box.

Surface Area of a Cylinder

Imagine a soup can. It has two circular ends and a body you can open and lay flat. The flattened body is a rectangle whose height is the same height as the can. The base of the rectangle s equal to the circumference of the can.

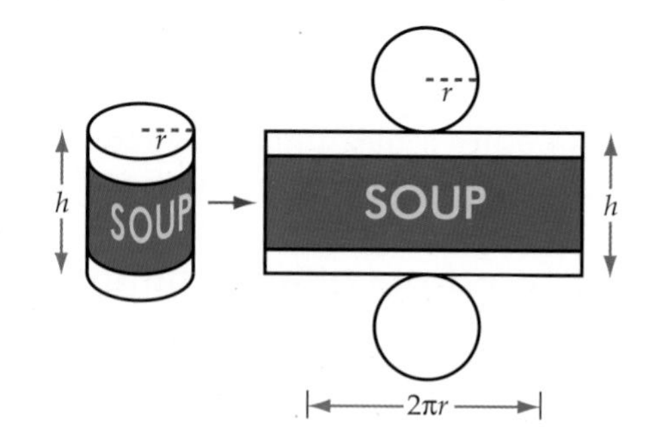

The total surface area of the cylinder is the sum of the area of the two circular ends and the area of the rectangle.

Find the surface area of the can if the radius of the cylinder is r and its height is h:

Surface area of 2 circles: 2 • area of circle $= 2 \bullet \pi r^2 = 2\pi r^2$
Surface area of rectangle: h • circumference $= 2\pi rh$
Total surface area of can: $= 2\pi r^2 + 2\pi rh$

This is the total area of the two circles that form the top and bottom of the can and of the rectangle, which is the body of the can rolled out flat.

Surface Area of a Sphere

You can also find the surface area of a solid that does not have any flat surfaces. The surface area of a sphere is directly related to its radius by the formula:

$$S = 4\pi r^2$$

Example

The surface area of this sphere is calculated as:

$$S = 4 \bullet \pi \bullet 10^2 = 400\pi \text{ cm}^2$$

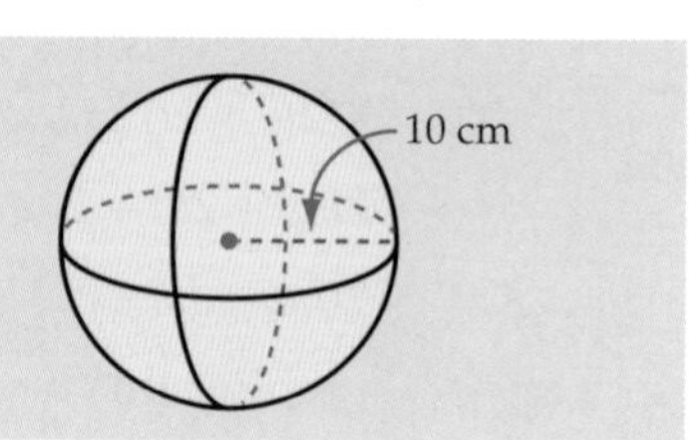

Think about surface area in terms of paint. A good measure of the surface area of a box or can or sphere is the amount of paint needed to cover these objects to be a uniform thickness.

Unit Conversion of Area

English Units

Area measurements can be converted from square feet to square inches using the fact that there are 144 square inches in 1 square foot. ($12 \text{ in} \bullet 12 \text{ in} = 144 \text{ in}^2$)

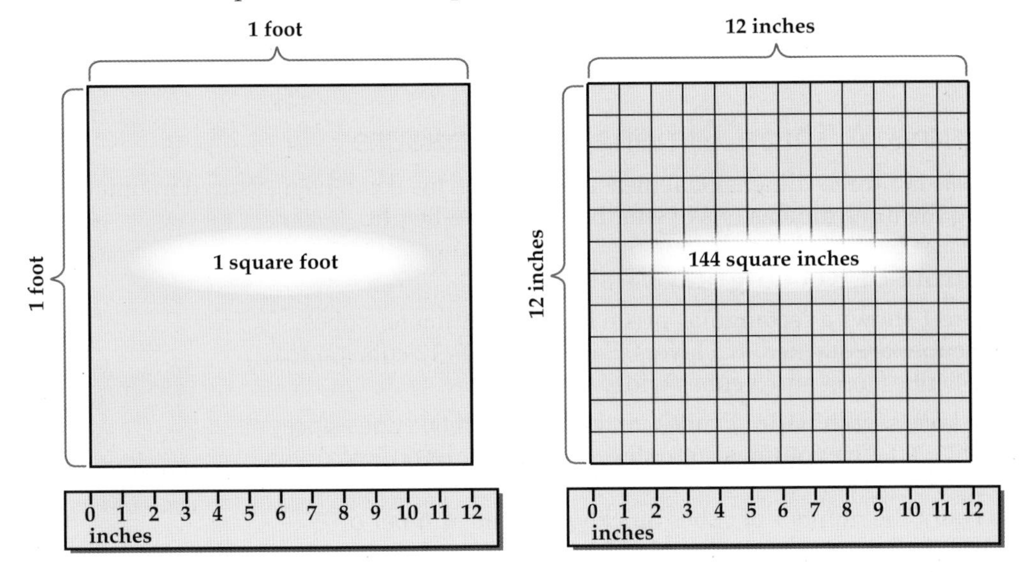

Example

Since 1 ft^2 = 144 in^2, then 5.5 ft^2 = 5.5 • 144 in^2 = 792 in^2.

Metric Units

Area measurements can also be converted from square millimeters (mm^2) to square centimeters (cm^2) and to square meters (m^2):

$$100 \text{ (or } 10^2\text{) } mm^2 = 1 \ cm^2$$

$$10{,}000 \text{ (or } 100^2\text{) } cm^2 = 1 \ m^2$$

$$1{,}000{,}000 \text{ (or } 1{,}000^2\text{) } m^2 = 1 \ km^2$$

Volume: A Three-Dimensional Measure

Volume is a three-dimensional measure. This means it can be measured in unit cubes, using the three dimensions of length, width, and height.

Volume of Rectangular Prisms

Figure 1 shows a rectangular prism built of centimeter cubes. The cubes are stacked in 5 layers. Each layer has 12 cubes since it is 4 cubes long and 3 cubes wide.

Five layers, each with 12 cubes, makes a total of 60 cubes since 5 • 12 = 60. The volume is 60 cubic centimeters, or 60 cm^3.

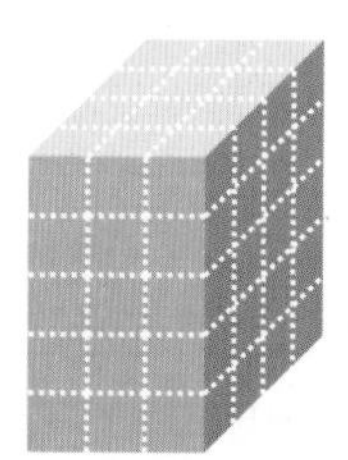

Figure 1

To count the number of cubes quickly, find the number of cubes in each layer and multiply by the number of layers. Since multiplication is commutative, there is more than one way to compute the volume of a rectangular prism.

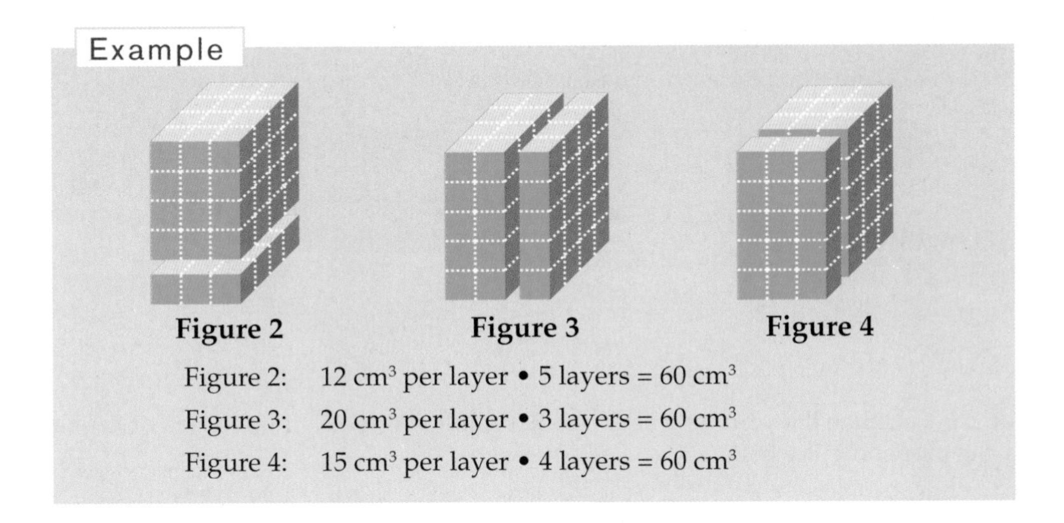

Figure 2: 12 cm^3 per layer • 5 layers = 60 cm^3

Figure 3: 20 cm^3 per layer • 3 layers = 60 cm^3

Figure 4: 15 cm^3 per layer • 4 layers = 60 cm^3

Counting the cubes gives the same result as using the formula for the volume, V, of a rectangular prism.

$$V = \text{length} \bullet \text{width} \bullet \text{height}$$
$$V = lwh$$

Whether you count cubes or measure each of the three dimensions with a ruler, you are multiplying the length (in centimeters) by the width (in centimeters) by the height (in centimeters) to get the volume (in cubic centimeters).

This volume formula can also be written: $V = \textit{base area} \bullet \textit{height}$ $\qquad V = Ah$

Volume of Parallel Solids

Parallel solids are three-dimensional objects that have congruent cross-sections.

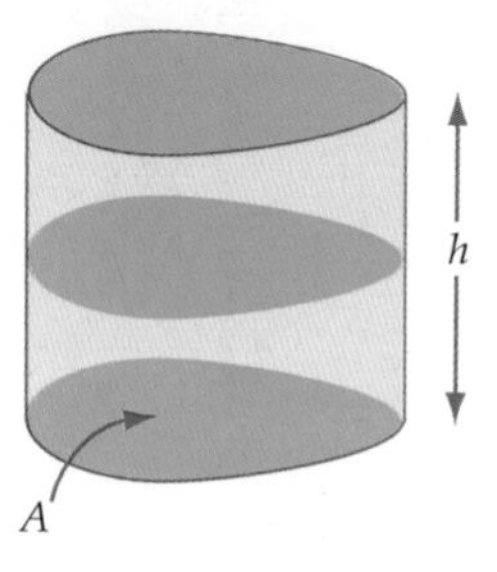

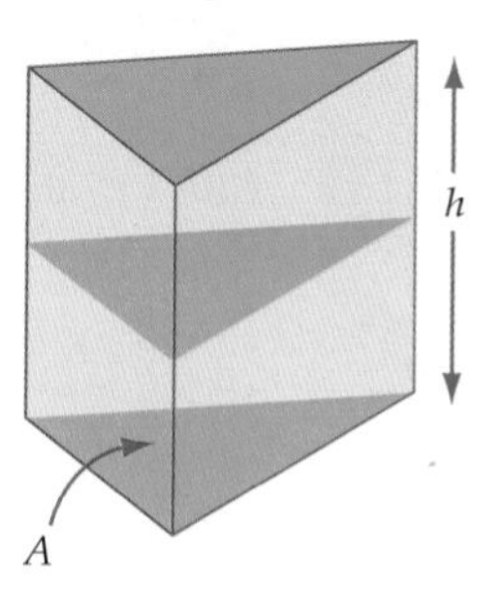

You can calculate the volume of any parallel solid by multiplying the area of the base by the perpendicular height.

Volume = base area • height $V = Ah$

Example

Calculate the volume of a cylinder with a radius of 3 inches and a height of 10 inches.

$V = Ah = \pi r^2 h$

$= \pi \bullet 3^2 \bullet 10 = 90\pi$

$\approx 3.14 \bullet 90 \approx 282.6 \text{ in}^3$

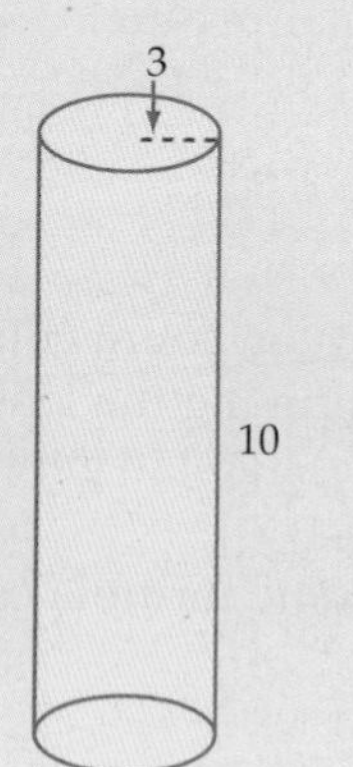

Volume of Point Solids

A *point solid* is a shape in which all faces or surfaces other than the base meet at a single point called the *apex*. The height of a point solid is the perpendicular distance from the apex to the base. Cross-sections taken parallel to the base are similar in shape to the base but smaller.

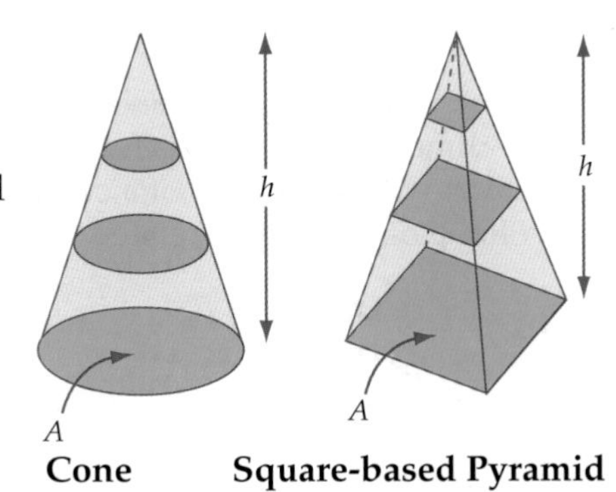

A *tetrahedron* is a point solid whose four faces are triangles. A *regular tetrahedron* has four faces that are equilateral triangles. For any tetrahedron, you can calculate its volume using any of the four faces as the base.

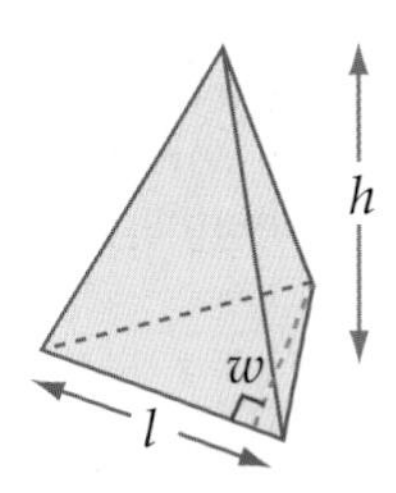

You can calculate the volume of any point solid by multiplying one-third of the base area times the height.

$$Volume = \frac{1}{3} \bullet \textit{base area} \bullet \textit{height} \qquad V = \frac{1}{3}Ah$$

Example

This point solid, a rectangular-based pyramid, has a height of 7 m and a base that is 3 m by 4 m.

$V = \frac{1}{3}Ah$

$V = \frac{1}{3} \bullet (3 \bullet 4) \bullet 7 = \frac{1}{3} \bullet 12 \bullet 7$

$= \frac{1}{3} \bullet 84 = 28 \text{ m}^3$

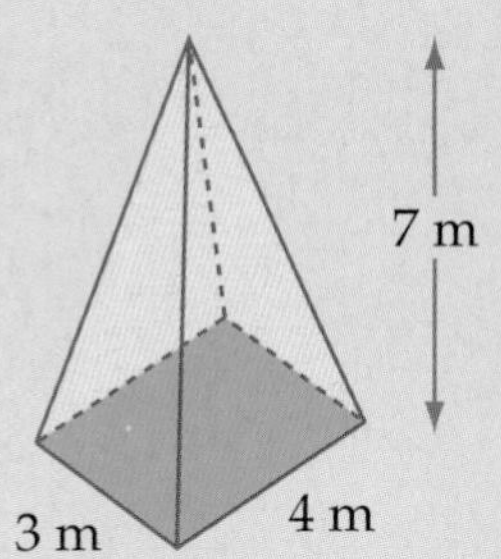

Volume of a Sphere

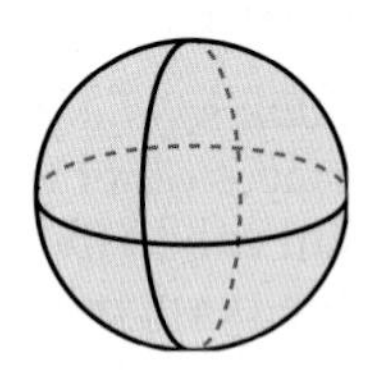

If a circle is rotated around one of its diameters through space, a three-dimensional object is formed. This is a *sphere*.
As with the circumference of a circle and the area of a circle, the volume of a sphere is related to its radius using π (pi).

In any sphere, the ratio of the volume, V, to the cube of its radius, r^3, is four-thirds of π.

$$\frac{\text{Volume}}{\text{radius cubed}} = \frac{4}{3}\pi \qquad V = \frac{4}{3}\pi r^3$$

The volume involves cubing a linear measure, the radius.

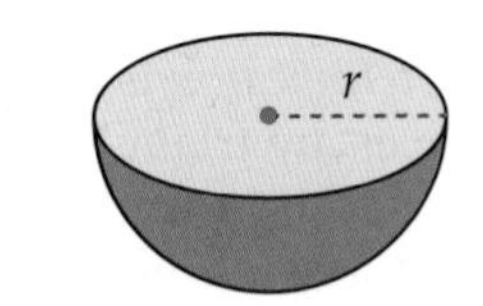

A hemisphere is a half of a sphere. The volume of a hemisphere is half the volume of a sphere with the same radius.
So, for a hemisphere with radius r:

$$V = \frac{1}{2}\left(\frac{4}{3}\pi r^3\right)$$

Example

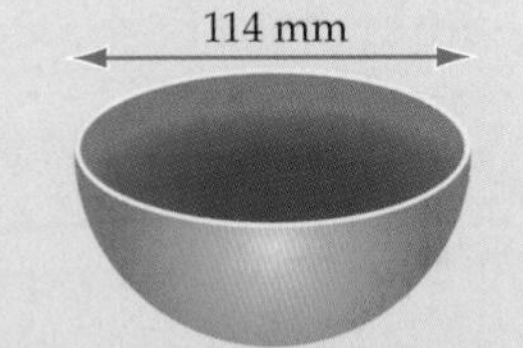

This bowl is in the shape of a hemisphere with a diameter of 114 mm. Its volume is half the volume of a sphere with a similar radius.

The radius is half of 114 mm, or 57 mm.

Compute the volume:

$$V = \frac{1}{2}\left(\frac{4}{3}\pi \bullet 57^3\right)$$

$$\approx \frac{1}{2}\left(\frac{4}{3} \bullet 3.14 \bullet 57^3\right)$$

$$= \frac{1}{2}(775{,}341.36)$$

$$= 387{,}670.68 \text{ mm}^3$$

GEOMETRIC SIMILARITY

CHAPTER 15

Similar Figures and the Similarity Ratio

Plane figures and solids that are mathematically *similar* to each other have the same shape but not necessarily the same size.

For two plane figures to be similar, they must satisfy these two conditions of similarity:

- The measures of corresponding angles are equal.
- The ratios of the lengths of every pair of corresponding sides are equal.

This ratio of the lengths of corresponding sides is called the *similarity ratio*, *k*.

Example

This diagram shows two similar quadrilaterals that have been moved into positions of similarity, with the center of similarity at point *O*.

The symbol for "is similar to" is ~, so you can write $ABCD \sim A'B'C'D'$.

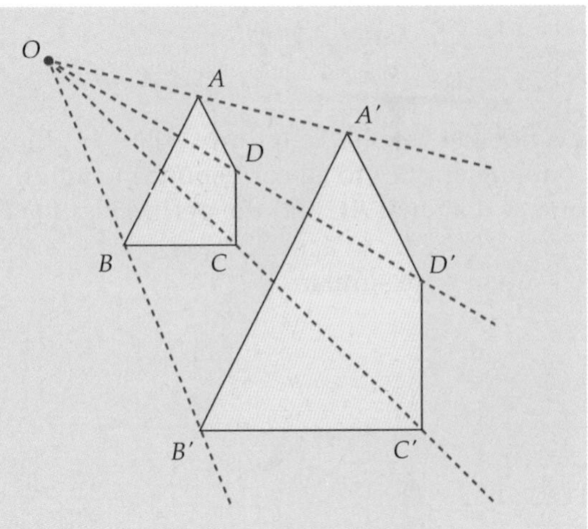

These quadrilaterals are said to be in the *position of similarity*.

There is more than one way to state a similarity ratio.

Example

For these figures, the similarity ratio, k, is either

$2:1 = 2$ for
$A'B'C'D' : ABCD$
It is the expansion that turns the smaller figure into the larger one: $A'B' = 2AB$, etc.

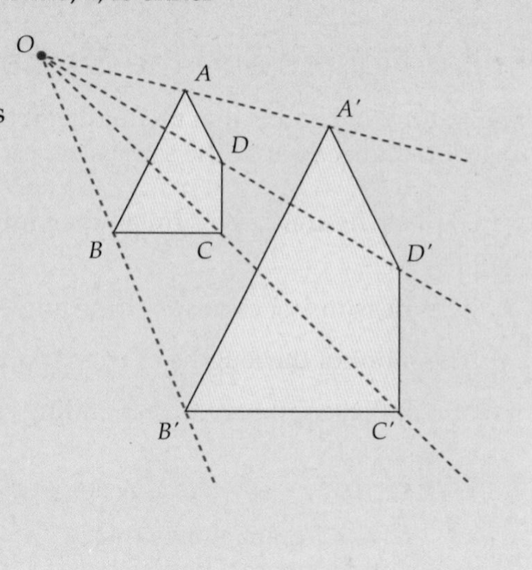

Or

$1:2 = \frac{1}{2}$ for

$ABCD : A'B'C'D'$
It is the shrinking in size that turns the larger figure into the smaller one:

$AB = \frac{1}{2}\ A'B'$, etc.

It is not just the side lengths of $A'B'C'D'$, that are double those of $ABCD$. The doubling applies to all corresponding length measures. So diagonal $A'C'$ is twice as long as diagonal AC and the perimeter of $A'B'C'D'$ is double the perimeter of $ABCD$.

All squares are similar:

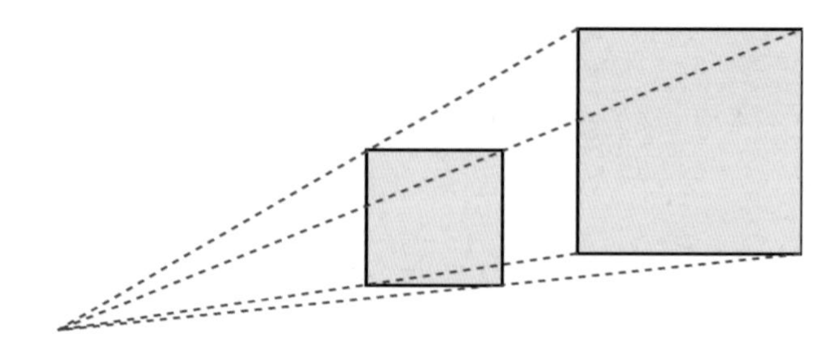

But not all rectangles are similar. The ratio of the pairs of corresponding sides of any two rectangles are not always equal.

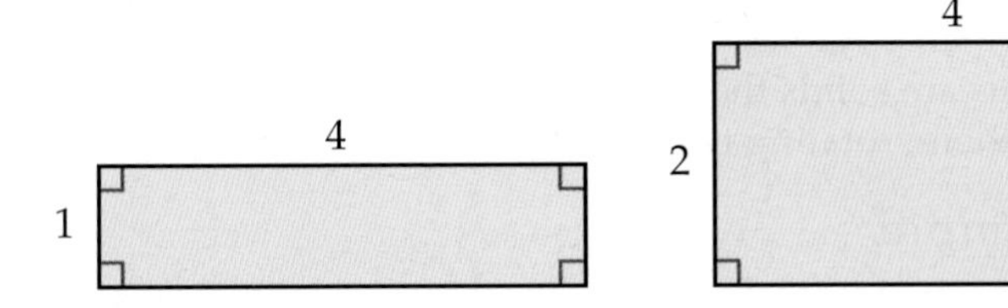

Hence, $\frac{4}{4} \neq \frac{2}{1}$.

And not all rhombuses are similar. Remember, a rhombus is a type of quadrilateral with two pairs of parallel sides and four equal sides. These two diagrams show that corresponding angles are not always equal.

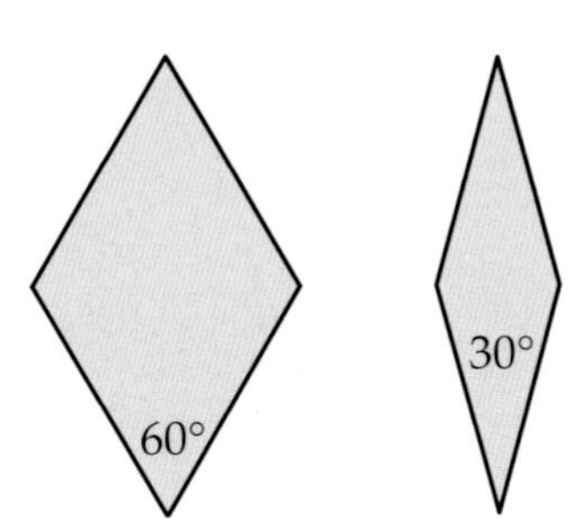

All circles are similar.

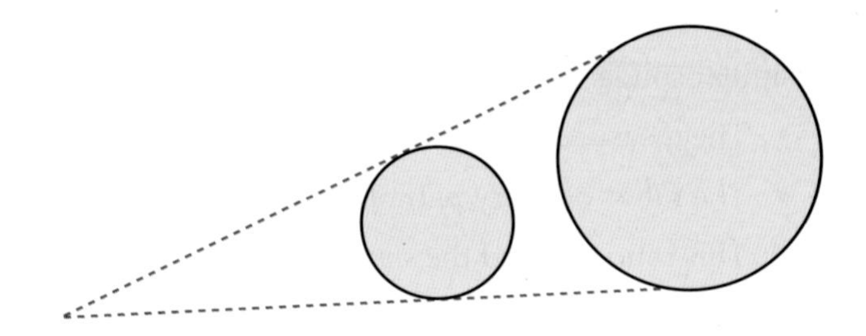

But not all sectors (parts of circles) are similar. You can see from these two examples of sectors that corresponding angles are not always equal.

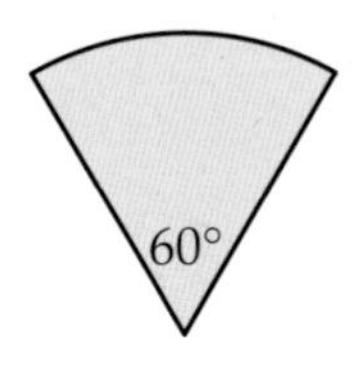

Similar Solids and the Similarity Ratio

Similar solids are solids that can be moved into positions of similarity by expanding or shrinking them, rotating them, or turning them around.

Example

Only three of the four hand shapes below are similar to each other.

The shape that is not similar is figure 4, the right hand. Moving this solid shape around will not put it into a position of similarity to any of the left hands.

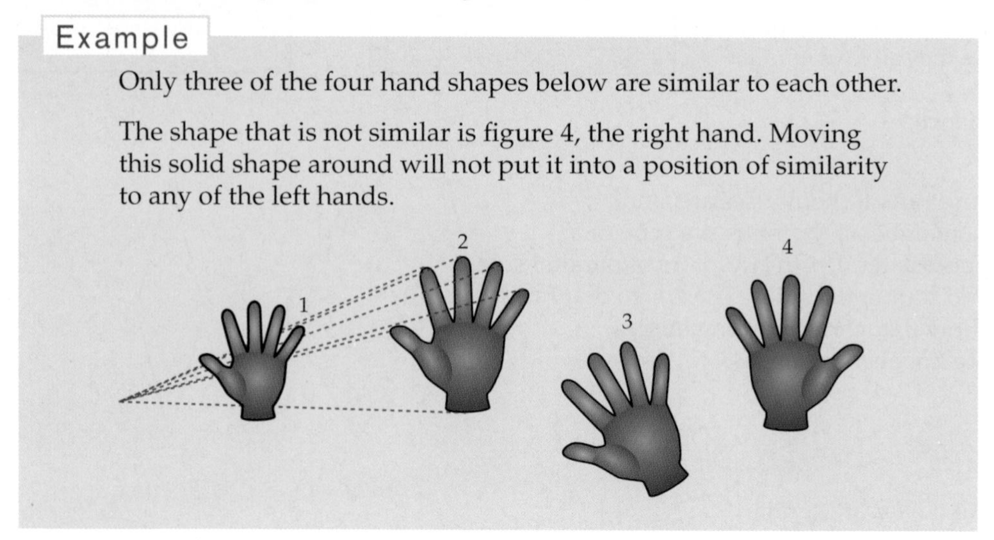

In similar solids:

- The measures of corresponding angles are equal.
- The ratios of every pair of corresponding length measurements are equal.
- This ratio is called the *similarity ratio*, k.

All cubes are similar.

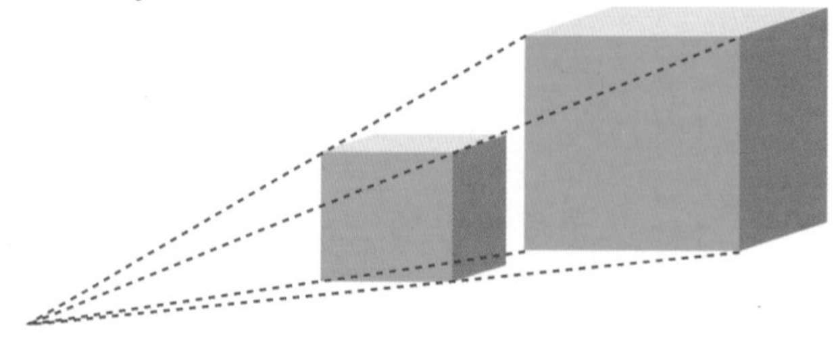

All spheres are similar.

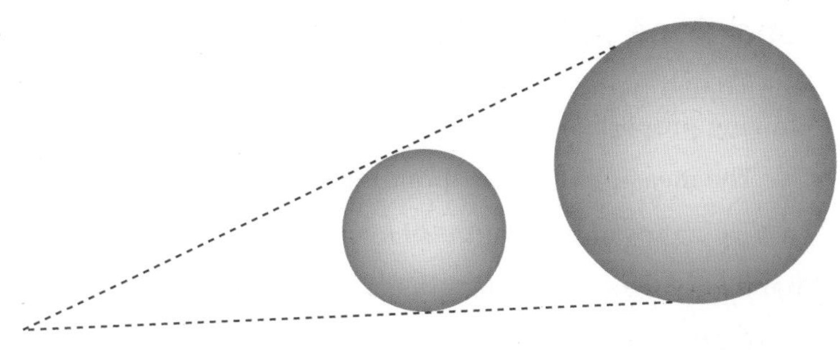

Not all rectangular prisms are similar.

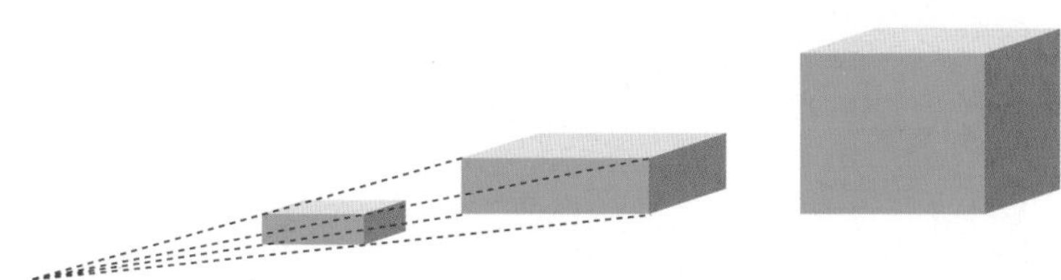

These are similar to each other.........but not to this.

Not all cylinders are similar.

Here, $\frac{d_1}{h_1} = \frac{d_2}{h_2} \neq \frac{d_3}{h_3}$

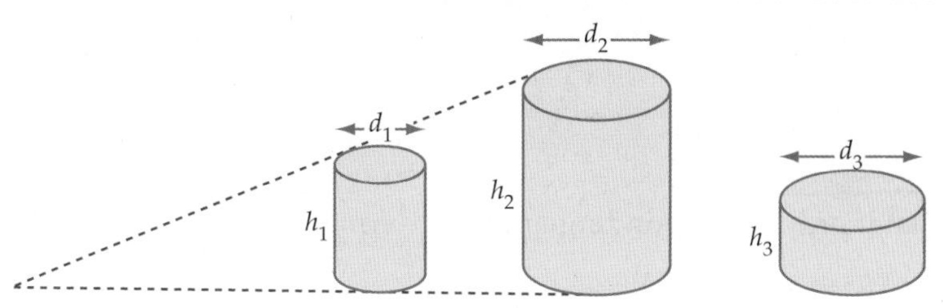

These are similar to each other........but not to this.

And not all cones are similar.

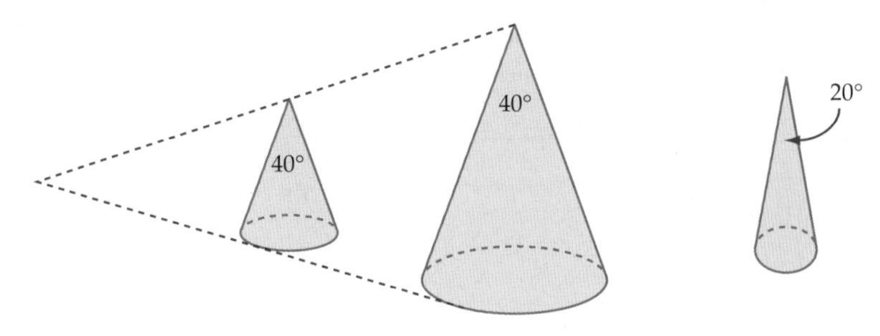

These are similar to each other........but not to this.

Conditions for Similar Polygons

To prove that two polygons are similar, you must show that they have both of the following properties:

- The measures of corresponding angles are equal.
- The ratios of the lengths of every pair of corresponding sides are equal.

What if only the ratios of lengths of corresponding sides are equal?

Example

These two figures are not similar. Corresponding side lengths are in the ratio 1 : 2, but the corresponding angles are not equal.

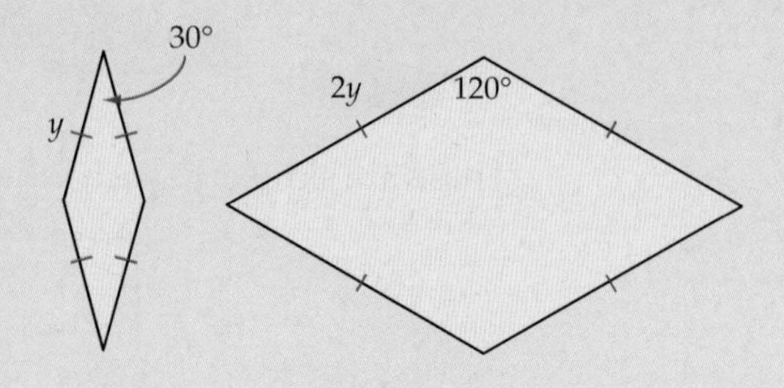

What if only corresponding angles are equal?

Example

These two figures are not similar. Corresponding angles all equal 120°, but corresponding side lengths are not in the same ratio. For four pairs of corresponding sides the ratio is 1 : 1; for the other two corresponding sides it is 1 : 2.

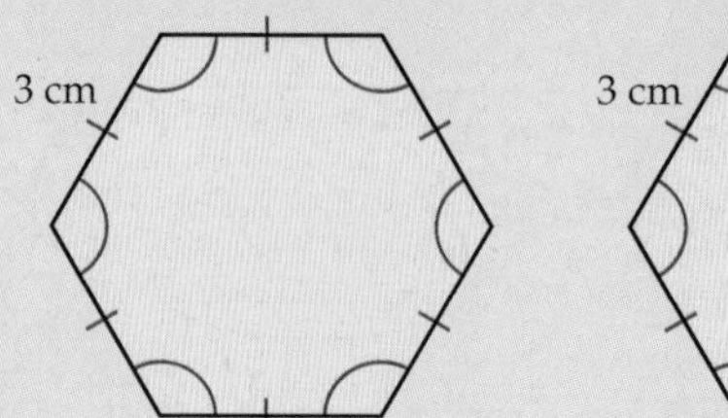

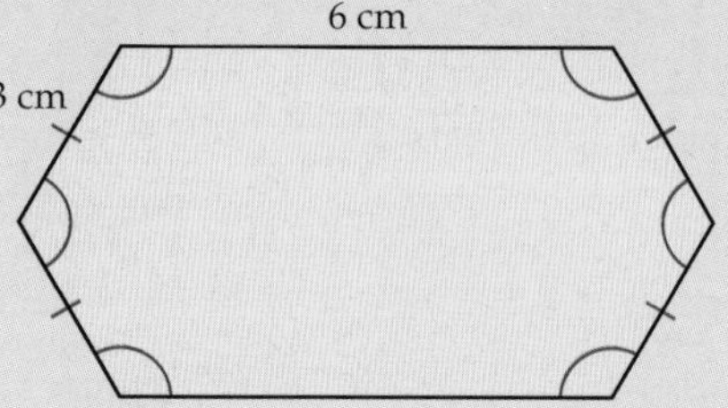

A Special Case: Conditions for Similar Triangles

Because triangles have fewer sides and angles, you can prove similarity with fewer facts. There are three ways in which you can decide if two triangles are similar.

1. If the corresponding angles are equal, then the triangles are similar. This will be true if just two pairs of corresponding angles are given as equal, since the third angle is found by subtraction from 180°.

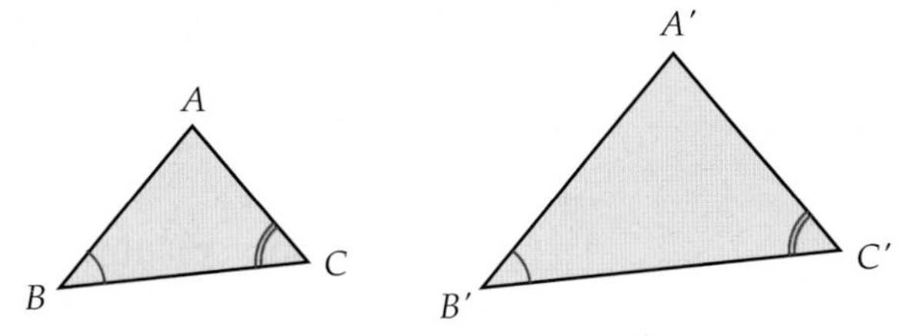

2. If the ratios of the lengths of the three pairs of corresponding sides are equal, then the triangles are similar.

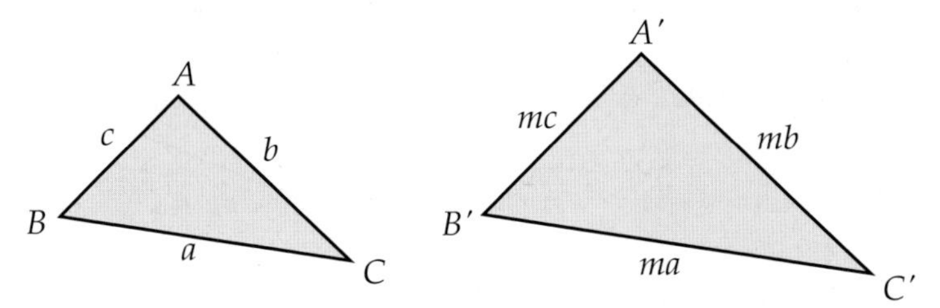

3. If the ratios of the lengths of two pairs of corresponding sides are equal, and the corresponding angles between them are equal, then the triangles are similar.

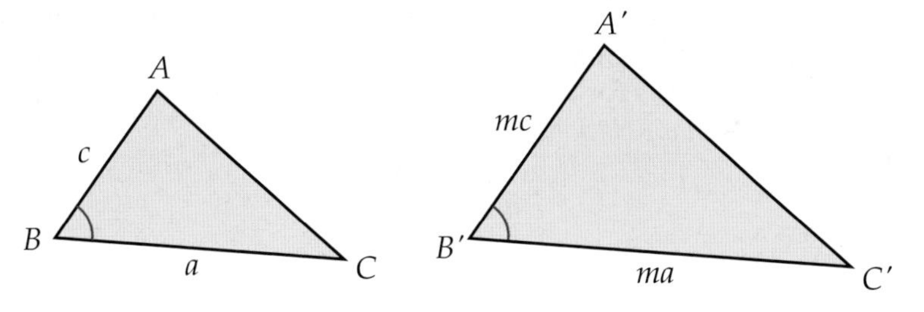

Finding Unknown Angles in Similar Figures

If two polygons are known to be similar, then their corresponding angles are equal. This means that if an angle in one shape is known, then the corresponding angle in the similar shape can be found by direct comparison.

Example

In the similar quadrilaterals below, $\angle A$ corresponds to $\angle E$, so the measures of these two angles are equal. Therefore, x measures 37°.

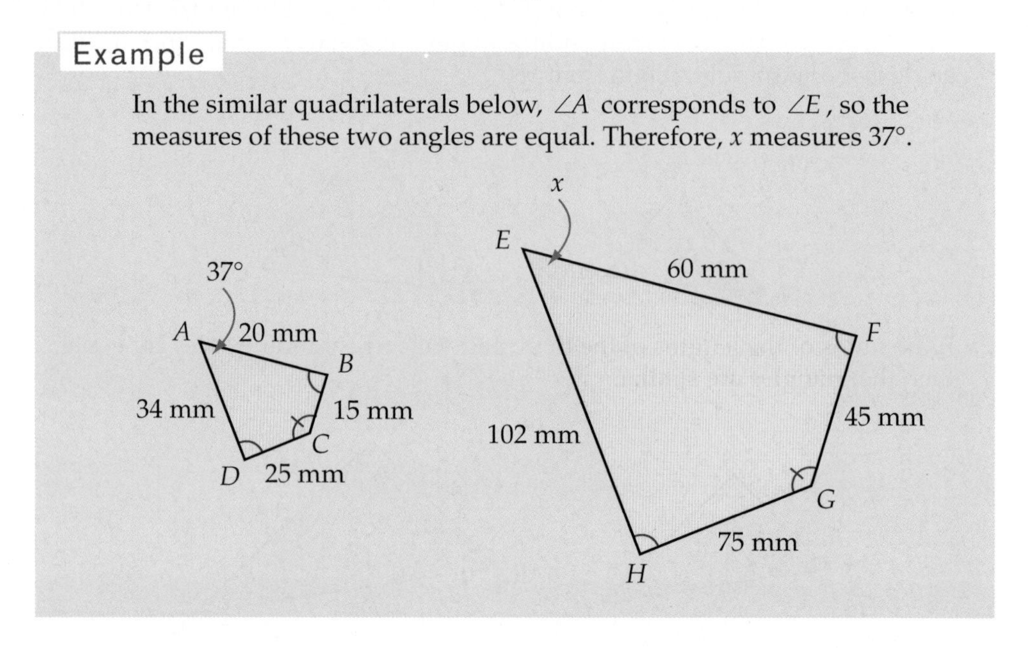

Finding Unknown Lengths in Similar Figures

If two triangles are similar, then ratios of all pairs of corresponding angles are equal, and can be expressed as a similarity ratio. Unknown lengths can be found using this ratio. Three common methods for doing this are explained on pages 263–266. (See also Chapter 19, *Proportional Relationships*.)

1. **Equivalent Ratios between Triangles**

In some problems, it is easy to determine the unknown lengths. Comparing the corresponding side lengths of the similar triangles gives the similarity ratio.

Example

Here, the similarity ratio equals 3, because

$12 : 4 = 3$

x must be three times the length of its corresponding side.

Thus, $x = 3 \bullet 7 = 21$ mm.

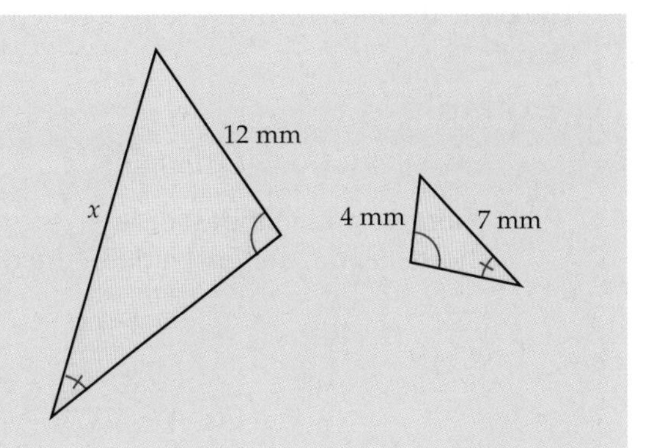

Example

These triangles are similar, so you can write the equivalent ratios between the two triangles.

$y : 19 = 10 : 30.5$

This ratio can be expressed as equivalent fractions.

$$\frac{y}{19} = \frac{10}{30.5}$$

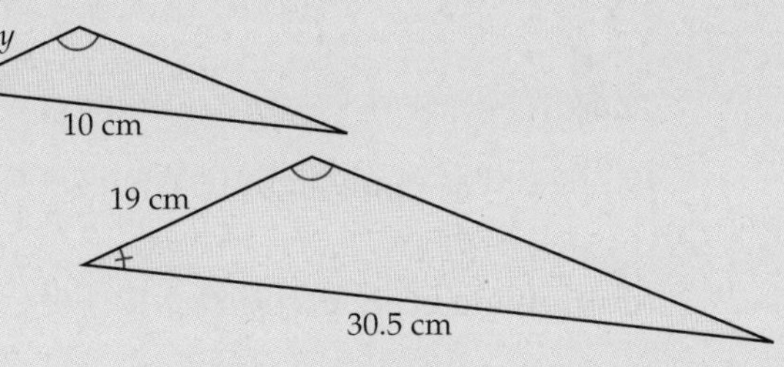

To find the value of y, multiply both of the equivalent fractions by 19. (The multiplication property of equality allows you to do this.)

$$\frac{y}{19} \bullet 19 = y = \frac{10}{30.5} \bullet 19 \approx 6.23$$

2. **Equivalent Ratios within Triangles**

The previous problem can also be solved by comparing ratios of side lengths within each triangle.

The basic theorem of ratios says, "Given two similar triangles, if $a : b = m : n$, then $a : m = b : n$."

Example

Given two similar triangles:

Shortest side : longest side in one triangle = shortest side : longest side in the other triangle.

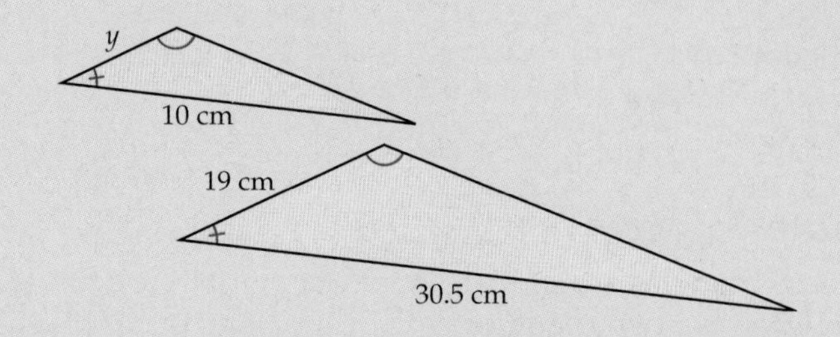

This gives: $y : 10 = 19 : 30.5$

Expressed as equivalent fractions: $\dfrac{y}{10} = \dfrac{19}{30.5}$

Multiplying each fraction by 10, the multiplicative inverse of $\dfrac{1}{10}$, gives: $\dfrac{y}{10} \bullet 10 = \dfrac{19}{30.5} \bullet 10$

Simplifying both sides gives: $y \approx 6.23$ cm

3. **Triangles in Positions of Similarity**

In this diagram, the parallel lines mean that $\Delta PRS \sim \Delta PQT$, with point P at the center of similarity and with a ratio of similarity of

$$k = \frac{PS}{PT} = \frac{9}{3} = 3$$

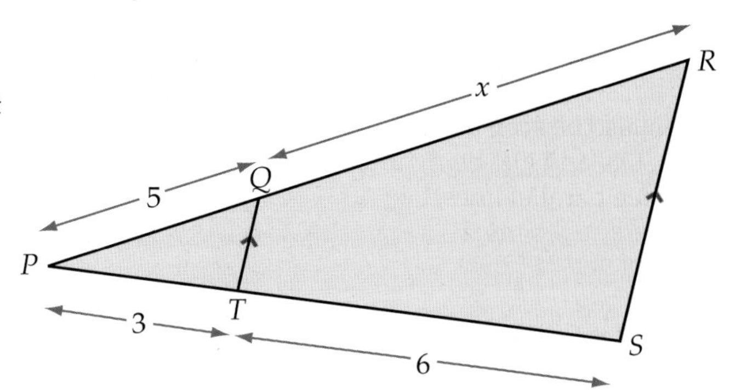

In fact, with the triangles in the position of similarity as shown here, the ratio of the two segments along one triangle is equal to the ratio of the two segments along the other triangle.

$$\frac{PT}{TS} = \frac{PQ}{QR}$$

Example

The ratio of the two segments along PR is equal to the ratio of the two segments along PS. This gives $PQ : QR = PT : TS$.

Substituting values: $5 : x = 3 : 6$

As equivalent fractions, with the letter in the numerator: $\frac{x}{5} = \frac{6}{3}$

Multiplying each fraction by 5: $\frac{x}{5} \bullet 5 = \frac{6}{3} \bullet 5$

Simplifying both sides: $x = \frac{6}{3} \bullet 5 = 10$

This diagram has three triangles in positions of similarity.

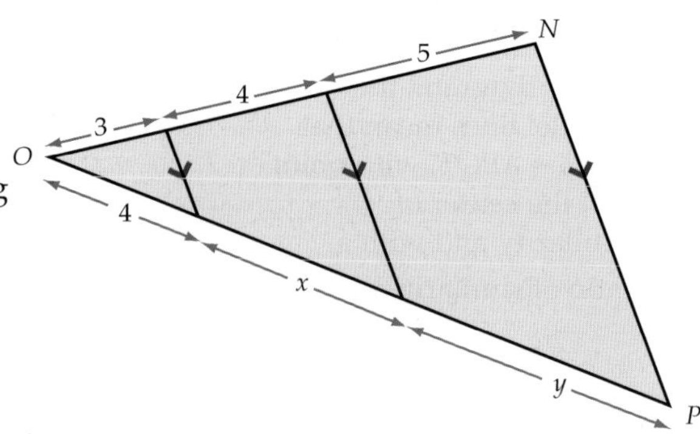

It can also be seen as two intersecting lines, *ON* and *OP*, being intercepted by three parallel lines.

A ratio table like the one below can be used to solve for x and y.

On $\overline{OP}$	4	$4 + x$	$4 + x + y$	x	y
On $\overline{ON}$	3	7	12	4	5

The ratio table shows five pairs of numbers that are all in the same ratio of 4 : 3. Use these equal ratios to solve for x and y.

Example

To find x, start with $\frac{x}{4} = \frac{4}{3}$.

Multiply both sides by 4.

$$\frac{x}{4} \bullet 4 = \frac{4}{3} \bullet 4$$

The solution is $x = \frac{16}{3} = 5\frac{1}{3}$.

To find y, start with $\frac{y}{5} = \frac{4}{3}$.

Multiply both sides by 5.

$$\frac{y}{5} \bullet 5 = \frac{4}{3} \bullet 5$$

The solution is $y = \frac{20}{3} = 6\frac{2}{3}$.

Similar Right Triangles

Many practical problems can be solved by knowing the acute angles and side lengths in a right triangle, particularly in the case where one of the side lengths is 1.

Example

These figures are similar right triangles. The similarity factor is k = 1,000, the length (in feet) of the string that is holding the kite.

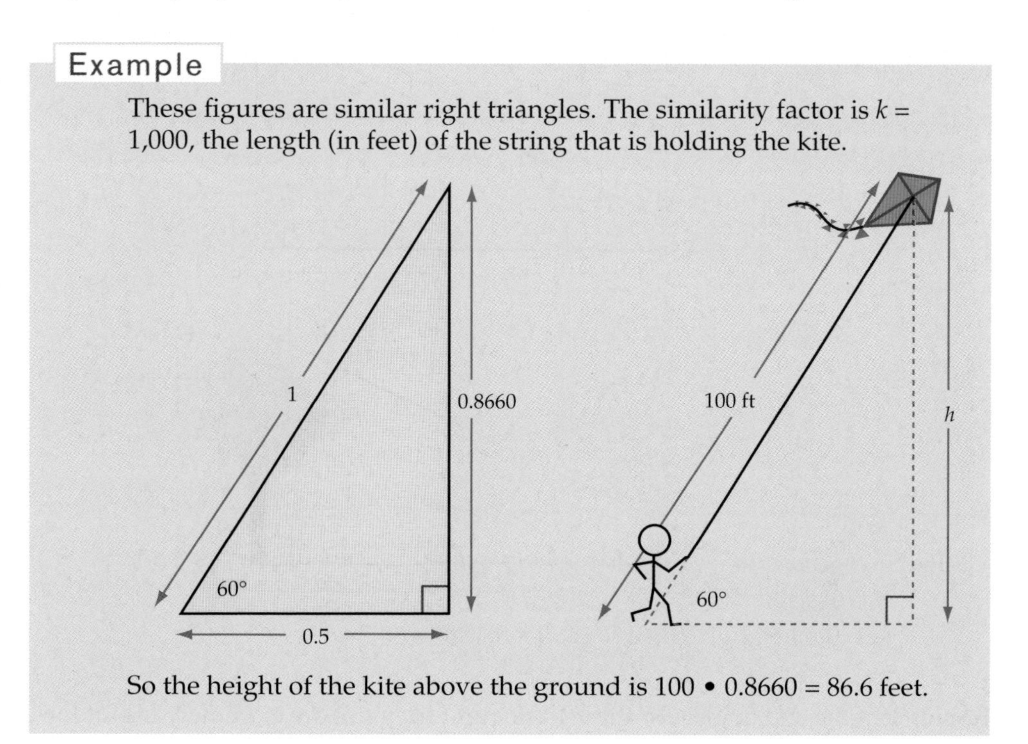

So the height of the kite above the ground is 100 • 0.8660 = 86.6 feet.

Example

These figures are similar right triangles. It is easy to see that the similarity factor is $k = 50$, the distance in feet to the base of the tree.

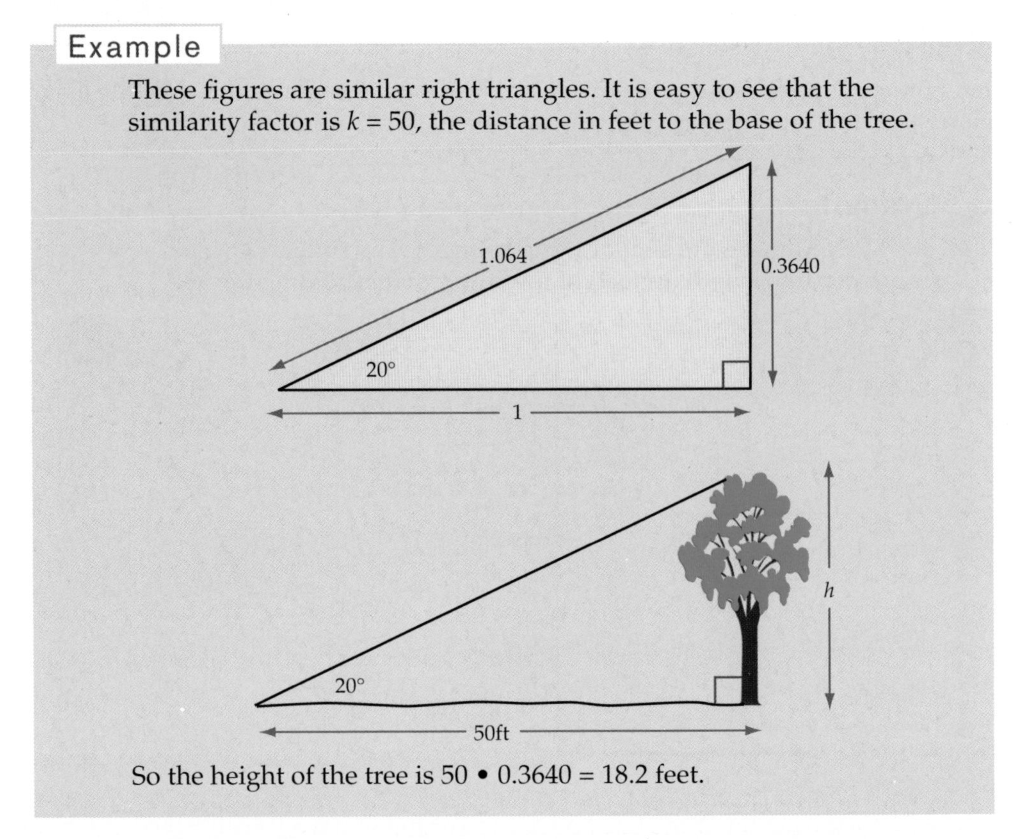

So the height of the tree is 50 • 0.3640 = 18.2 feet.

The side lengths and angles for these basic right triangles are extremely useful for solving practical problems of these types.

Areas of Similar Shapes

For similar shapes, the ratio of corresponding areas is equal to k^2, the square of the ratio of similarity.

Example

In this figure, the large triangle is similar to the small shaded one, with a similarity ratio of $k = 3$, so the ratio of their areas is $k^2 : 1 = 9 : 1$.

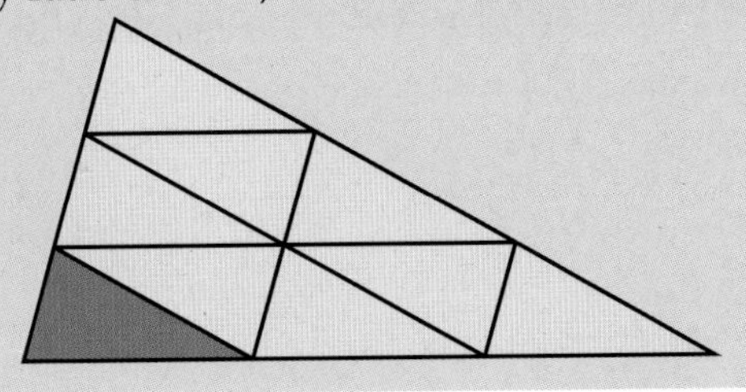

Example

In this figure, the two cylinders are similar, with a similarity ratio of $k = \frac{h_2}{h_1} = \frac{d_2}{d_1} = 4$. Therefore, the ratio of their areas is $k^2 : 1 = 16 : 1$.

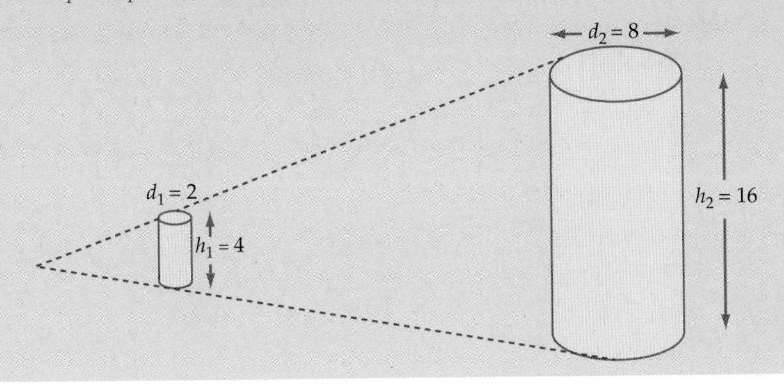

Once you have found that the surface area of the small cylinder is 10π, then you can multiply this answer by 16 to get the surface area of the large cylinder, 160π.

Volumes of Similar Solids

For similar solids, the ratio of corresponding volumes is equal to k^3, the cube of the ratio of similarity.

Example

In this figure, the two cylinders are similar, with a similarity ratio of $k = 4$. Therefore, the ratio of their volumes is $k^3 : 1 = 64 : 1$.

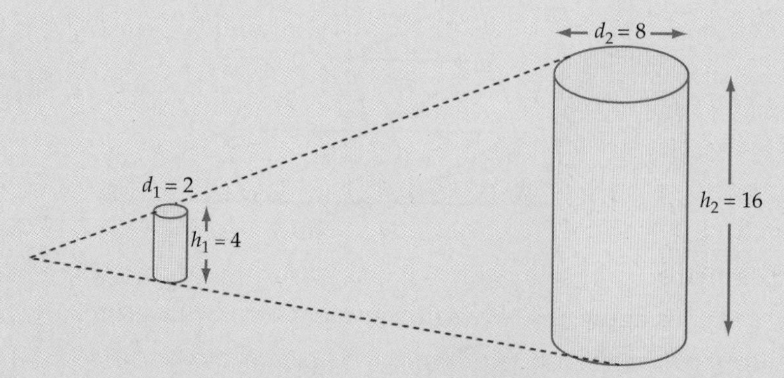

Once you have shown that the volume of the small cylinder is 4π, then you can multiply this answer by 64 to get the volume of the large cylinder, 256π.

UNITS AND QUANTITIES

CHAPTER 16

What Is a Unit?

How tall are you? How long is the pencil on the teacher's desk? How much does the dog weigh? How long is a class period at your school? The answers to questions like these are expressed in units.

Example

John is 64 inches tall. The pencil is $5\frac{1}{4}$ inches long.

The dog weighs $21\frac{1}{2}$ pounds.

Class periods at Jefferson School are 50 minutes long.

Inches, pounds, and minutes are examples of units.

Units give you a standard. Every inch is exactly the same length as every other inch. Every pound weighs the same as every other pound. Each minute lasts the same amount of time as every other minute. When you measure something, you compare it to standard units. When you measure length, you compare the length of what you are measuring to the units on a ruler. Units let people communicate about length, weight, duration, and anything else that can be measured or counted.

When you add or subtract, the units remain the same; add inches to inches and you still have inches. When you multiply or divide, you create a new unit from the units used.

Example

Divide miles by hours to get a unit for measuring speed: $\frac{\text{miles}}{\text{hour}}$.

Multiply people by hours to get a measure of an amount of work, people-hours.

Chapter 16

What Is a Quantity?

Look at these two statements:

1. If 63 is divided by 7, the result is 9.
2. If John bicycles 63 miles in 7 hours, his average speed is 9 miles per hour.

Each statement uses the same numbers, 63, 7, and 9. But statement 1 just uses mathematics, while statement 2 applies mathematics to a situation in the physical world, using units of measurement: miles, hours, and miles per hour.

A *quantity* is a number with a unit of measurement. A quantity refers to something counted or measured. Statement 1 has no quantities, only numbers. Numbers without quantities are sometimes called *pure numbers*. Statement 2 has three quantities, each with different units of measurement: 63 miles, 7 hours, and 9 miles per hour.

The unit of measurement tells what standard was used as the measure, and the number tells how many of that unit were measured. In statement 2, 63 miles, 7 hours, and 9 miles per hour are all examples of quantities.

The word quantity can be also be used in a different way in mathematics. You use the word quantity in describing $a(b + 2)$ by saying a times the quantity $b + 2$. This is a different use of the word quantity. Both are correct.

Adding and Subtracting with Units

You can add and subtract quantities that have the same units. The results have the same units.

Example

3 apples + 5 apples = 8 apples

3 inches + 5 inches = 8 inches

$\frac{1}{2}$ hour − $\frac{1}{4}$ hour = $\frac{1}{4}$ hour

\$5.50 – \$.55 = \$4.95

Apples, hours, inches, and dollars (\$) are all units.

You cannot add or subtract unlike units.

Example

These are not possible to add or subtract:

3 apples + 5 inches $\frac{1}{2}$ hour – $.55

Multiplying Quantities

Multiply a Quantity by Another Quantity

When you multiply quantities, the units get multiplied along with the numbers. You can multiply unlike quantities. The units of the product are usually different from the units of either factor.

Example

3 people working 5 hours = 3 people • 5 hours = 15 people-hours

People-hours are a unit for measuring the amount of work done in terms of 1 person working 1 hour.

3 people working 5 hours is equivalent to 15 people-hours.

Example

A rectangle that is 3 inches wide by 5 inches tall has an area of 15 square inches. *Square inches* is a unit used to measure area.

A 3 × 5-inch rectangle has an area equivalent to 15 squares that are each 1 inch by 1 inch. Mathematically, you write:

$$3 \text{ in} \cdot 5 \text{ in} = 15 \text{ sq in} \quad \text{or}$$

$$3 \text{ in} \cdot 5 \text{ in} = 15 \text{ in}^2$$

Notice what happens when you multiply two quantities. Each quantity refers to something countable or measurable in the situation, such as, hours or inches.

After you multiply, you have a new unit that refers to something different in the situation: people-hours or square inches. This change in units does not happen when you add or subtract; it usually happens when you multiply or divide.

Multiply a Quantity by a Dimensionless Number

Sometimes you multiply a quantity with units by a number that has no units. In these situations, the units of the product and the factor with units are the same.

Example

John had 4 times as much money in his pocket as Millie.
Millie had $8.00.
How much money did John have?

Answer: John had 4 • $8.00 = $32.00.
The factor 4 has no units, so the product of 4 and 8 dollars is 32 dollars.

The product has the same units as one of the factors. Numbers with no units are called *dimensionless numbers*.

When you enlarge or shrink the size of something, you use a "scale factor." A scale factor is a common kind of dimensionless number.

Example

You may enlarge a photograph so that it is 1.5 times as large. The scale factor 1.5 is a ratio that compares lengths before and after the enlargement.

A photo is:	2 inches by 3 inches	(size before)
You increase by:	1.5	(scale factor)
New photo size:	(2 • 1.5) by (3 • 1.5)	(size after)
	3 inches × 4.5 inches	

Because a scale factor is dimensionless, the units of the product are the same as the original units.

Dividing Quantities

Divide a Quantity by a Quantity

Just as you can multiply unlike quantities (numbers with different units), you can also divide them. And just as with multiplication, the units of the quotient are usually different from the units being divided.

Example

Divide miles by hours and you get a new quantity, miles per hour or mph, which is a measure of speed.

Divide pounds by square inches, and you get a new quantity, pounds per square inch, which is a measure of pressure.

Divide dollars by hours and you get a new quantity, dollars per hour, which is a measure of value.

These examples show how common it is to create a rate from dividing two quantities.

In many situations, you multiply and divide several different units. The rules for operating with units are the same as the rules for operating with numbers and letters that represent numbers. When units appear by fractions, you "cancel" like terms (identity property).

Example

$\cancel{7} \bullet \left(\frac{3}{\cancel{7}}\right) = 3$ (Cancel the 7s.)

$2 \text{ } \cancel{\text{hours}} \bullet 15 \frac{\text{miles}}{\cancel{\text{hour}}} = 30 \text{ miles}$ (Cancel hours.)

$\frac{\$1.50}{\cancel{\text{lb}}} \bullet 2 \text{ } \cancel{\text{lbs}} = \$3.00$ (Cancel the pounds.)

Divide a Quantity by a Quantity with the Same Unit

When you divide like quantities, the quotient has no units. It is dimensionless.

Ratios with No Units

Example

These ratios have no units:

The relationship of the circumference to the diameter of a circle is π.
The shape of a rectangle indicated by the ratio of height to width.

Types of Quantities

There are four basic types of quantities:

- Quantities that use a count
- Quantities that use a unit of measurement
- Quantities that use a ratio of two quantities
- Quantities that use a rate formed from two quantities

Quantities that Use a Count

Example

There were 32 students in the class.

The box contained 32 marbles.

The bus made 32 stops before they got to the destination.

A regular polygon with 32 sides looks almost like a circle.

Maria got the right answer 32 times in a row.

In these statements, 32 students, 32 marbles, 32 stops, 32 sides, and 32 times are all examples of quantities that give a count of the number of objects in some situation.

Remember that a quantity consists of a number with a unit attached to it. In the quantity 32 students, the number is 32, and the unit is "students." The unit names the thing that is being counted. Almost anything you can refer to can be counted.

Quantities that Use a Unit of Measurement

Example

The bus had a length of 32 feet.

The blanket had an area of 32 square feet.

The fish tank had a volume of 32 gallons.

The power was out for a period of 32 hours.

The dog had a weight of 32 pounds.

The book had a value of 32 dollars.

Each of these statements involves the same number, 32, but in each case the number is associated with a different unit. 32 feet, 32 square feet, 32 gallons, 32 hours, 32 pounds, and 32 dollars are all examples of quantities.

The example shows not only the quantity but the *dimension*: the bus has a length, the blanket has an area, the fish tank has a volume, the power outage is measured in time, the dog has a weight, and the book has a value.

A *dimension* is a feature of a situation that can be measured. With any dimension, there are often different units that can be used. For example, length can be measured using inches, feet, and yards; or it could be measured with centimeters, meters, and kilometers.

If the dimension of the two quantities is the same, you can add or subtract them is you have the same units.

Example

1 inch + 1 centimeter =

You can convert 1 inch to 2.54 centimeters.

$$2.54 \text{ cm} + 1 \text{ cm} = 3.54 \text{ cm}$$

These units are in the same dimension, length, so you can add them after you convert them to the same unit.

Other dimensions can be derived from basic dimensions.

Example

$\frac{\text{length}}{\text{time}}$ is speed.

length • width is area.

$\frac{\text{weight}}{\text{area}}$ is pressure.

The units for many derived quantities are rates.

Example

Speed can be measured in $\frac{\text{miles}}{\text{hour}}$ or $\frac{\text{meters}}{\text{second}}$.

Pressure can be measured in $\frac{\text{lbs}}{\text{square inch}}$.

Sometimes the units are single units of measure, like miles, and sometimes units combine to make new units, like miles per hour.

Below is a list of single units of measure that are used in this course.

Dimension (what is measured)	Example of Units of Measurement	Words for What is Measured	Examples of Commonly Measured Items
length	inch, foot, yard, mile centimeter, meter, kilometer	length, width, height, depth, distance	pencil, skateboard, rope, bus, hallway, street
area	square inch, square foot, square mile square centimeter, square meter, square kilometer	area, surface area	blanket, table top, wall, field, city, state, country
volume	cubic inch, cubic foot, cubic mile cubic centimeter, cubic meter, gallon, liter	volume, capacity	marble, cup, basketball, fish tank, room, pool
time	second, minute, hour, day, year	time, period, duration	duration of a class, a meeting, a game, a vacation, life
weight	ounce, pound gram, kilogram	weight	letter, package, book, person, car
value	cents, dollars foreign currency	monetary value, amount, cost, price, wage	value of book, cost of house, hourly pay

English System of Measurement

John is 64 inches tall. You can also refer to his height as "5 feet, 4 inches." Feet and inches belong to the same system for measuring length.

The table below shows the relationships among units of length in the English System.

	Each column shows equivalent measures						
Inches	12	24	36				
Feet	1	2	3				5280
Yards	$\frac{1}{3}$	$\frac{2}{3}$	1	220	440	880	1760
Miles				$\frac{1}{8}$	$\frac{1}{4}$	$\frac{1}{2}$	1

All quantities in a single column are equivalent: 12 inches = 1 foot = $\frac{1}{3}$ yard.

Metric System of Measurement

Another system for measuring length is the metric system. The metric system is used almost everywhere in the world except in the United States. Even in the United States it is frequently used along with the customary English units. Scientists everywhere use the metric system.

The table below shows the relationships among units of length in the metric system.

	Each column shows equivalent measures					
Millimeter	10	100	1000			
Centimeter	1	10	100			
Meter	.01	.1	1	10	100	1000
Kilometer						1

Quantities that Use a Ratio

Quantities can involve a ratio of two other quantities. (For more on ratios, see Chapter 17, *Ratios*).

Like all quantities, a ratio quantity is given by a number and an associated unit. The unit of a ratio quantity is actually a ratio of units. The meaning of a ratio quantity is one quantity "per" a unit amount of another quantity.

A ratio expresses a relationship between two quantities. When you talk about a relationship, you refer to each quantity and say what kind of relationship exists between them.

Example

There are 520 students in Monroe School, and on a particular day 104 of the students are absent. You can use this information to find the ratio of absent students to all students:

$$\frac{\text{\# of absent students}}{\text{\# of students in the school}} = \frac{104}{520} = \frac{104 \bullet 1}{104 \bullet 5} = \frac{1}{5}$$

For every 5 students at Monroe, 1 student was absent.

You could also say:

1 student out of 5 students was absent.
20% of the students were absent.

Here is an example of solving a problem before thinking about the units.

Example

There are 2.5 times as many students at Midland School as at Lowland School. There are 256 students at Lowland.

You can use this information to find how many students are at Midland.

Two quantities are being compared: number of students at Midland and number of students at Lowland. The comparison is a ratio:

$$\frac{\text{Midland students}}{\text{Lowland students}} = 2.5$$

Using pure numbers, the answer is:

$$2.5 \bullet 256 = 640$$

If you examine the units, you see your calculation is correct and the units in your answer are 640 students at Midland:

$$\frac{\text{Midland students}}{\cancel{\text{Lowland students}}} \bullet (\cancel{\text{Lowland students}}) = (\text{Midland students})$$

Notice that the unit, Lowland students, appears in the denominator on the left and in the numerator on the right. So, you can cancel these units, just as if this were a fraction with digits. The units that remain on the left side after the cancellation are Midland students. This gives the unit in the product.

$$2.5\left(\frac{\text{Midland students}}{\text{Lowland students}}\right) \bullet 256\ (\text{Lowland students}) = 640\ (\text{Midland students})$$

Example

Using the same ratio, if you knew that there were 1000 students at Lowland, the ratio 2.5 tells are that there would be 2500 students at Midland.

Quantities that Use a Rate

Quantities can involve a rate formed from two other quantities. (For more on rates, see Chapter 18, *Rates*.)

Like all quantities, a rate quantity is given by a number and an associated unit. The unit of a *rate quantity* is actually a ratio of units, and the meaning of a rate quantity is one quantity "per" a unit amount of another quantity.

Rates and ratios are very closely connected. They both contain the word "per" linking the two quantities. Any quantity using the word "per" is a rate or ratio.

Example

A car traveled 270 miles in 6 hours. Use this information to find the average rate of speed:

$$\frac{\text{miles traveled}}{\text{hours spent}} = \frac{270}{6} = 45\left(\frac{\text{miles}}{\text{hours}}\right)$$

The average speed was 45 miles per hour.

Here is an example of solving a problem before thinking about the units.

Example

Lincoln School needs to have 3 computers per classroom. There are 41 classrooms. Use this information to find how many computers are needed at Lincoln School.

$$3 \bullet 41 = 123$$

So, the number of computers needed is 123.

You have performed the calculation $3 \bullet 41 = 123$. To show why the calculation is the correct, you can examine the units:

$$\left(\frac{\text{computers}}{\cancel{\text{classrooms}}}\right) \bullet (\cancel{\text{classrooms}}) = (\text{computers})$$

Notice that the unit, classrooms, appears in the denominator on the left and in the numerator on the right. You can cancel these units, just as if this were a fraction with digits. The units that remain on the left side after the cancellation are computers. This gives the units in the product.

You can also show the full calculation together with the units:

$$3\left[\frac{\text{computers}}{\cancel{\text{classrooms}}}\right] \bullet 41\ [\cancel{\text{classrooms}}] = 123\ [\text{computers}]$$

For each quantity, the number is given first and the units are given next in parentheses.

Converting Units

Example

Joe measures the length of a hallway in yards, but Mary wants the length of the same hallway in feet.

There are three feet in one yard, so you can write 3 [feet] = 1 [yard].

You can divide both sides by 1 [yard] to get $\frac{3[\text{feet}]}{1[\text{yard}]} = 1$.

This is a ratio of quantities, so it is a rate and can be written like this:

$$1 = 3\,\frac{[\text{feet}]}{[\text{yard}]} \text{ or } 1 = 3 \text{ feet/yard}$$

This is an example of a *conversion factor*. It can be read as "3 feet per yard."

Since $a \bullet 1 = a$ for any number, you can multiply any number or quantity by a conversion factor without changing its value.

Example

If your hallway measures 7 yards, you can multiply by the conversion factor and not change the value.

$$7\,[\text{yards}] \bullet 3\left[\frac{\text{feet}}{\text{yard}}\right] = 21\ \cancel{\text{yards}} \bullet \frac{\text{feet}}{\cancel{\text{yard}}}$$

$\frac{\text{yards}}{\text{yards}}$ cancel the yards and you get,

7 yards = 21 feet

Conversion Factors and Unit Rates

In the equation $\frac{3[\text{feet}]}{1[\text{yard}]} = 1$, the right side of the equation is 1. Thus, you can multiply any quantity, Q, by either side (1 or $\frac{3[\text{feet}]}{1[\text{yard}]}$) and not change the value of the quantity, Q.

The right side, 1, has no units. It is a *dimensionless quantity*. This means you can multiply any quantity, Q, by either 1 or $\frac{3[\text{feet}]}{1[\text{yard}]}$ and not change the *dimensions* of the quantity, Q.

You have found 7 yards = 21 feet. These quantities have the same dimension, *length*.

Yards to Inches

You can also relate inches to feet in an equation similar to the yards-to-feet example:

$$1 = 12\left[\frac{\text{inches}}{\text{foot}}\right]$$ The conversion factor is 12 inches per foot.

And you know that $1 = 3\left[\frac{\text{feet}}{\text{yard}}\right]$

Since you can multiply both sides of an equation by 1, you can get:

$$1 \bullet 1 = 3\left[\frac{\text{feet}}{\text{yard}}\right] \bullet 12\left[\frac{\text{inches}}{\text{foot}}\right]$$

Carrying out the multiplications on each side, you get a new conversion factor, this time from yards to inches:

$\left[\frac{\text{feet}}{\text{foot}}\right]$ cancel the feet and you get:

$$1 = 36\left[\frac{\text{inches}}{\text{yard}}\right]$$ The conversion factor is 36 inches per yard.

Using Unit Analysis to Solve Problems

By paying attention to the units, you can solve problems.

Example

Jack's car went 120 miles on 3.5 gallons of gas. Using gas at the same rate, how much will it cost to go 50 miles if it costs $25.00 to fill an 11-gallon tank?

What is the situation?

It would be useful to make some tables and graphs to help you see this situation. Another approach is to hide the numbers and variables and just work with the units. This is called a *unit analysis.*

You figure out which operations (such as addition, subtraction, multiplication, and division) you need to perform on the given units to end with the units you want for the answer. You often multiply and divide to cancel out unwanted units. When you get the units you want for the answer, you go back and put the numbers in and do the same calculations with the numbers.

If you keep thinking about what the units refer to in the problem situation, you will understand how what you are doing makes sense.

What will the units of the answer be?

The question asks for the cost to go 50 miles. The answer will be a cost in dollars:

$ for a trip of 50 miles

What units are given in the situation?

Example

Jack's car went 120 miles on 3.5 gallons of gas. Using gas at the same rate, how much will it cost to go 50 miles if it costs $25.00 to fill an 11-gallon tank?

You have: $\frac{\text{miles}}{\text{gallon}}$ for a trip

miles you want to travel

$\frac{\$}{\text{gallon}}$ for a tank

You want to find the $ for a 50-mile trip.

What quantity should you start with?

Usually in units analysis, it doesn't matter where you start. Start where it makes sense to you. You can start with the quantity that has $ in it:

$$\frac{\$}{\text{gallon}}$$

Your goal is to end up with $. So you want to cancel out the gallon from the denominator. You need a gallon from another quantity in the numerator to cancel the gallon in the denominator. You have $\frac{\text{miles}}{\text{gallon}}$, but that gallon is also in the denominator.

What operation should you use?

You cannot add or subtract unlike units.

If you multiplied: $\frac{\$}{\text{gallon}} \bullet \frac{\text{miles}}{\text{gallon}}$, you would get $\left[\frac{\$ \bullet \text{miles}}{\text{gallon} \bullet \text{gallon}}\right]$. That does not work.

If you divide:

$$\left(\frac{\frac{\$}{\text{gallon}}}{\frac{\text{miles}}{\text{gallon}}}\right) = \frac{\$}{\text{miles}}$$

That equation makes sense. But you have \$ per mile and you want your answer in \$.

So, you need to get the miles canceled from the denominator.

If you look back at the units given in the situation, the quantity you have not used is "miles." And miles times \$ per mile will get you to \$.

$$\frac{\$}{\text{miles}} \bullet \text{miles} = \$$$

Once you have worked out the units, go back and put the numbers in the equations.

$$\frac{\text{miles}}{\text{gallon}} \text{ for a trip} = \frac{120}{3.5}; \qquad \frac{\$}{\text{gallon}} \text{ for a tank} = \frac{25}{11}; \qquad \text{miles to travel} = 50$$

Example

$$\left(\frac{\frac{\$}{\text{gallon}}}{\frac{\text{miles}}{\text{gallon}}}\right) \text{ or } \frac{\left(\frac{25}{11}\right)}{\left(\frac{120}{3.5}\right)} = \left(\frac{25}{11}\right) \bullet \left(\frac{3.5}{120}\right) = \frac{35}{528} \text{ (\$ per mile)}$$

$$\frac{\$}{\text{miles}} \bullet \text{miles} = \frac{35}{528} \bullet 50 = \frac{875}{264} \approx 3.314$$

It costs approximately \$3.31 to drive 50 miles in Jack's car.

Summary: $\left(\frac{\frac{\$}{\text{gallon}}}{\frac{\text{miles}}{\text{gallon}}}\right) \bullet \text{miles} = \$$

RATIO

CHAPTER 17

Ratio and Proportion

In school mathematics, the subjects of "ratio" and "proportion" are usually discussed together because the idea of proportion is an extension of the idea of ratio. This chapter will focus almost entirely on ratio. A later chapter (19) will discuss the idea of proportion and its relationship to ratios.

Ratio Is a Comparison by Division

A *ratio* is a simple idea that is used in mathematics and in everyday life.

A ratio is the *comparison* of two quantities by *division*.

Example

Suppose that 24 students and 6 teachers go on a field trip.

You can compare the number of students to the number of teachers by dividing.

Since $24 \div 6 = 4$, you can say that there are 4 times as many students as teachers. You can also say there are 4 students to every teacher. A way of saying this using ratio is to say, "The ratio of students to teachers is 4 to 1."

Comparisons by Division versus Comparisons by Subtraction

A ratio is a comparison by division, but you can also compare by *subtraction*. Division and subtraction give quite different kinds of comparisons.

When you compare two quantities by division, it is called a *relative* comparison. A relative comparison tells you something about one of the quantities in terms of the other quantity.

On the other hand, a comparison by subtraction produces an *absolute* comparison. An absolute comparison tells you the *difference* between the two quantities.

Comparing two numbers, a and b, by subtraction gives a new number:

the *difference* $(a - b)$

Comparing two numbers, a to b, *where* $b \neq 0$, by division gives a new number:

the *ratio*, $\frac{a}{b}$

Relative comparison and absolute comparison are both useful, but they serve different purposes. There are times when a relative comparison is more appropriate than an absolute one, and vice versa.

Example

Here are the populations of two cities in the years 1990 and 2000.

	1990	2000
City *A*	4,000	7,000
City *B*	10,000	15,000

Comparison by Subtraction:

City A: 7,000 – 4,000 = 3,000
City B: 15,000 –10,000 = 5,000
City B grew by 2,000 more people than City A in the 10 years.

Comparison by Division:

City A: $\frac{7{,}000 - 4{,}000}{4{,}000}$ = 0.75 or 75 percent over the 10 years

City B: $\frac{15{,}000 - 10{,}000}{10{,}000}$ = 0.66 or 66 percent over the 10 years

City A grew 75% while City B grew 66% over the 10 years.

Relative Comparisons

There is a wide variety of ways to talk about any relative comparison, so it is worth taking time to examine this language.

Example

Suppose that 24 students and 6 teachers go on a field trip.

A relative comparison of the number of students to the number of teachers:

To make this comparison, divide the number of students by the number of teachers:

$$\frac{24}{6} = \frac{4}{1} = 4$$

Here are ways of talking about the student-teacher ratio:

There are 4 times as many students as teachers.

There are 4 students for every teacher.

There are 4 students per teacher.

There is a 4 to 1 ratio of students to teachers.

The ratio of students to teachers is 4 : 1.

A relative comparison of the number of teachers to the number of students:

To make this comparison, divide the number of teachers by the number of students:

$$\frac{6}{24} = \frac{1}{4} = 0.25$$

Here are ways of talking about the teacher-student ratio:

There are 0.25 times as many teachers as students.

There are $\frac{1}{4}$ as many teachers as students.

There is 0.25 of a teacher for every student.

There is a 1 : 4 ratio of teachers to students.

The ratio of teachers to students is 1 to 4.

There is 1 teacher for every 4 students.

Both of these comparisons are called "part-part" comparisons, since the teachers and the students are each part of the whole field trip group, and these comparisons are between these two parts (the number of students and the number of teachers). Another type of comparison is "part-whole."

You could determine several part-whole comparisons from this same situation.

A part-whole comparison of the number of teachers to the number of people on the field trip:

To make this comparison, divide the number of teachers (6) by the total number of people (24 students + 6 teachers):

$$\frac{6}{30} = \frac{1}{5} = 0.2$$

Here are ways of talking about this ratio:

0.2 of the people on the field trip are teachers.

The proportion of the people who are teachers is 0.2

20% of the people are teachers.

1 out of (every) 5 people is a teacher.

$\frac{1}{5}$ of the people are teachers.

The fraction of the people on the field trip who are teachers is $\frac{1}{5}$.

The ratio of teachers to the total number of people on the field trip is 1 to 5.

The ratio of teachers to the total number of people on the field trip is 1 : 5.

A part-whole comparison of the number of students to the number of people:

To make this comparison, divide the number of students by the total number of people on the field trip:

$$\frac{24}{30} = \frac{4}{5} = 0.8$$

Here are ways of talking about this ratio:

0.8 of the people on the field trip are students.

The proportion of the people who are students is 0.8.

80% of the people are students.

4 out of (every) 5 people are students.

$\frac{4}{5}$ of the people are students.

The fraction of the people on the field trip who are students is $\frac{4}{5}$.

The ratio of students to the number of people on the field trip is 4 to 5.

The ratio of students to the number of people on the field trip is 4 : 5.

Bar Diagrams

Bar diagrams are a way of representing and solving problems involving part-part and part-whole ratios.

Example

Lowland College has 650 students and teachers. Midland College has 40% more than Lowland. Both schools have 12 times as many students as teachers. How many students are there at Midland?

Find the Total Population of Students and Teachers at Midland

For Lowland, you know that 100% = 650 students and teachers. Draw a bar with 10 units, each representing 10% of the total. Each unit represents 65 people.

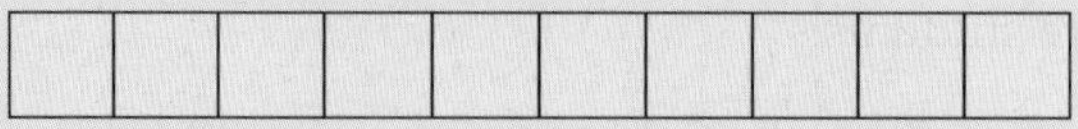

Each unit represents 10% of 650 students + teachers = 65

Since Midland has 40% more students and teachers, use the same bar but add 4 units. 14 units represents 140% of the people at Lowland.

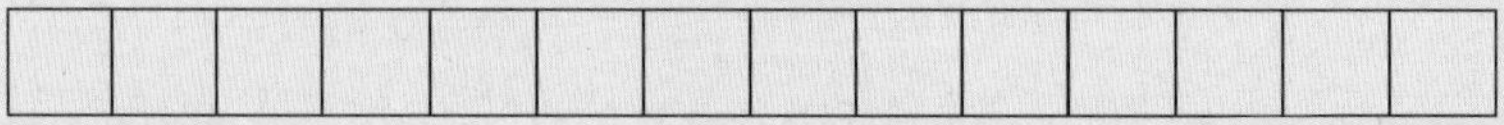

Students + teachers at Midland: $14 \bullet 65 = 910$

Find the Number of Students at Midland

Now you know that Midland has 910 people. Use a new bar diagram to find the number of *students only* at Midland.

Number of students = 12 • number of teachers.
If one unit is teachers then 12 units would be students.
Divide the bar into 13 units. Each unit represents 70 people because there are 910 people total ÷ 13 bars.

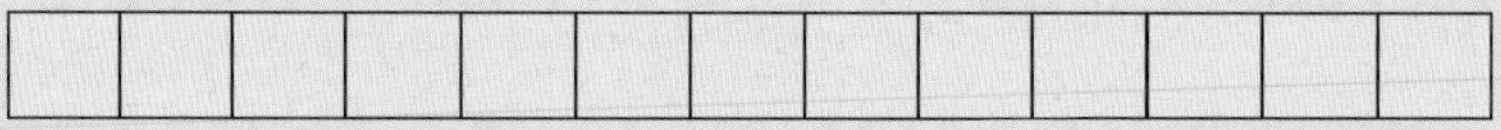

Unit Ratios

Unit ratios are ratios written as *some number* to 1. They provide an easy way to make comparisons between different ratios.

Example

A cooking club has 25 girls and 15 boys. A gymnastics club has 18 girls and 10 boys. Which club has the larger ratio of girls to boys?

The ratio of girls to boys in the cooking club is:

$$25 : 15 = 5 : 3$$

As a unit ratio, this is $\frac{5}{3} : 1 = 1.\overline{6} : 1$

The ratio of girls to boys in the gymnastics club is:

$$18 : 10 = 9 : 5$$

As a unit ratio, this is $\frac{9}{5} : 1 = 1.8 : 1$

Comparing the unit ratios shows the gymnastics club has the larger ratio of girls to boys.

Example

A 2.5 pound pack of lamb chops is priced at $9. A 1.6 pound pack of steak is priced at $5.60. Which meat is cheaper – lamb chops or steak?

The ratio of price to weight for the lamb chops is:

$$9 : 2.5 = 18 : 5 = 3.6 : 1$$

This unit ratio is called a unit price.
The unit price for lamb chops is $3.60 per pound.

The ratio of price to weight for the steak is:

$$5.6 : 1.6 = 56 : 16 = 7 : 2 = 3.5 : 1$$

The unit price for steak is $3.50 per pound.

Comparing these unit prices you see that the price of steak is slightly less than the price for lamb chops

Division Connects Ratios to Other Concepts

The concept of "division" is central to the idea of ratio. Division also plays a role in these six other concepts, but none of them is the same as ratio:

quotient	percent
rational number	rate
fraction	per

Quotients and Ratios

The term *quotient* is used in two related, but quite different, ways.

On the one hand, a quotient is the "whole number" part of the result of dividing one number by another. Any part less than the divisor left over in the division is the "remainder."

Example

The result of the division 23 ÷ 4 can be stated as the quotient 5, with remainder 3.

In this usage, a quotient is always an integer. This definition tends to be the one used in early school mathematics and also in higher-level courses that discuss the system of integers formally.

In the second usage, a "quotient" is the result of dividing one number by another.

Example

The result of dividing the "dividend" 23 by the "divisor" 4 is the quotient 5.75.

This is the way the word "quotient" is used in applications. The quotient, when used in this way, is a ratio of the quantity to 1.

Ratio and quotient both involve division. But, there is a difference in the way the two words are typically used. A ratio expresses the relationship of one quantity to another with each of the quantities having units, such as 4 to 1 is the ratio of students to teachers. A quotient, on the other hand, is simply a term for the result of dividing one number by another. A quotient does not need to have any units.

Rational Numbers and Ratios

A rational number is the result of dividing one integer by another.
Or, in other words, a rational number is the quotient of two integers.

Example

1.25 is a rational number since it equals $5 \div 4$.
Similarly, 1.3333 … is a rational number since it is equivalent to $4 \div 3$.

So, a rational number can be described as the ratio of two integers. *Whole number ratios* are those ratios where the numbers in the ratio are both whole numbers.

While the ratio of any two integers always gives a rational number, not every ratio gives a rational number. For instance, the ratio of the circumference of a circle to its diameter is π to 1, or just π. The number π is not a rational number; it is an irrational number. An irrational number is one that cannot be written as the ratio of two integers.

Fractions and Ratios

A fraction is a form of representing a number using a fraction bar.

Example

$\frac{3}{4}$ is a fraction.
The fraction bar indicates the division $\frac{3}{4}$ and represents the number $3 \div 4$. To find the decimal form of a fraction, simply carry out the division, either by hand or with a calculator:

$$\frac{3}{4} = 0.75$$

A ratio can be expressed as a fraction.

Example

If there are 21 girls and 28 boys in a class, the ratio of girls to boys is 3 to 4, or 0.75 to 1. As already noted, sometimes a ratio is not indicated in this typical "___ to ___" form, but as a single number:

The girl-boy ratio is 0.75.

The girl-boy ratio is 3 : 4 or 3/4 or $\frac{3}{4}$.

Still, there is a difference between a ratio and a fraction. A ratio is a quantity and, as such, typically has *units*. Specifically, a ratio expresses the relationship of one quantity to another, such as girls to boys in the above example. A fraction, on the other hand, is simply a way of indicating the division of one number by another. A fraction does not need to have any units.

You have discussed "fractions" as a particular way of representing a number such as $\frac{3}{4}$. But "fraction" is also an ordinary word in English that means much the same as the word *portion* or *proportion*. For example, "What fraction of the pie is mine?" The word "fraction" here means "part of," "portion," or "proportion."

Fractions are often used to describe part-whole relationships. However, fractions are not the only way to represent part-whole relationships. Each of the following is a valid way of describing a part-whole relationship:

$\frac{3}{4}$ of the students passed the test.

0.75 of the students passed the test.

75% of the students passed the test.

Also, a number represented as a fraction does not necessarily refer to part of a whole.

Example

The slope of a line is $\frac{3}{4}$.

Here is a fraction that is not expressing a part-whole relationship. Rather, it is the ratio of the rise to the run. So, although this is a fraction, it is not comparing parts to a whole.

Percents and Ratios

To illustrate the link between percents and ratios, return to the example where 6 teachers and 24 students went on a field trip. You know that 30 people went on the field trip. You can compute the ratio of teachers to the total number of people:

$$6 \div 30 = 0.2$$

A number such as 0.2 can always be expressed as a ratio of some number to 100.

In this case, $0.2 = \frac{20}{100}$ and the ratio is 20 to 100.

A ratio of something to the special denominator 100 is called a percent. So you can say that 20% of the people on the field trip are teachers. This means:

$$\frac{\text{teachers}}{\text{people on the field trip}} = \frac{6}{30} = \frac{20}{100}$$

These two statements are equivalent:

The ratio of teachers to all people on the field trip is 20 to 100.

20% of the people on the field trip are teachers.

Ratio Quantities

These table gives common examples of ratios expressed in ordinary language, and the number and unit of the ratio quantity. For each ratio quantity, the middle column of the table shows the number, and the column at the right shows the ratio of units.

	Example Expressing a Ratio Quantity	Ratio as Number	New Unit
1.	20% of the students in the school were absent that day.	0.2	$\frac{\text{absent students}}{\text{students in the school}}$
2.	There are 2.5 times as many students at Midland as at Lowland.	2.5	$\frac{\text{Midland students}}{\text{Lowland students}}$
3.	In just one year, the student population increased by a factor of 1.4.	1.4	$\frac{\text{students this year}}{\text{students last year}}$
4.	A rectangle with length about 1.6 times its width is called a golden rectangle.	1.6	$\frac{\text{length}}{\text{width}}$
5.	The circumference of a circle is π times its diameter.	π	$\frac{\text{circumference}}{\text{diameter}}$
6.	The copy machine gives enlargements that are 150% of the original size.	1.5	$\frac{\text{enlarged size}}{\text{original size}}$
7.	The slope of the road was 0.15.	0.15	$\frac{\text{rise of a section of road}}{\text{run of a section of road}}$
8.	John took 3 times as long as Pierre to finish the assignment.	3	$\frac{\text{John's time}}{\text{Pierre's time}}$
9.	Drying the fruit decreased its weight by 40%.	0.4	$\frac{\text{decreased weight}}{\text{original weight}}$
10.	The sales tax was 8.5%.	0.085	$\frac{\text{tax on item}}{\text{listed price of item}}$

Examples 1 to 3 involve a ratio of counts.
Examples 4 to 7 involve a ratio of length measurements.
Example 8 involves a ratio of time intervals.
Example 9 involves a ratio of weights.
Example 10 involves a ratio of monetary values.

Rates, Ratios, and "Per"

The words *rate* and *ratio* are often used in similar ways. Specificially, the word rate is often used to compare quantities that have different units. A quantity is a number with units, and a ratio of two quantities is another quantity consisting of a number with units. In the field trip example, where there are 24 students and 6 teachers, you can say that the ratio of students to teachers is 4 to 1.

You can express the same idea by saying, "There are 4 students per teacher." Here you have taken the pair of numbers from the ratio, 4 to 1, and expressed them as a single number: 4 students per teacher. This usage with "per" is often called a "rate."

This table shows examples of quantities that have a rate formed from quantities measured in different units. For each rate quantity, the middle column of the table shows the number and the column at the right shows the ratio of units.

	Example Expressing a Rate Quantity	Number	Unit
1.	There are 3 computers per classroom.	3	$\frac{\text{computers}}{\text{classroom}}$
2.	The car's average speed on the trip was 45 miles per hour.	45	$\frac{\text{miles}}{\text{hour}}$
3.	When you bought pencils last year, you paid only $0.20 per pencil.	0.20	$\frac{\$}{\text{\# of pencils}}$
4.	Today I had to pay $0.30 per pound for potatoes.	0.30	$\frac{\text{dollars}}{\text{pound}}$

Examples 3 and 4 are *unit prices*. Each is a rate of dollars per unit amount of what you are buying.

CHAPTER 18

RATES

Rates and Ratios

A ratio where the units being compared are different is sometimes called a *rate*.

Example

These are rates:

miles per hour (a rate of speed)

people per square mile (a measure of population density)

As in all ratios, calculating a rate involves comparing two quantities by division.

Example

If you travel 20 miles in 4 hours, you can compare these two quantities by dividing the distance by the time:

$$\frac{20 \text{ miles}}{4 \text{ hours}} = \frac{5 \text{ miles}}{1 \text{ hour}} = 5 \text{ miles per hour or } 5 \text{ mi/hr or } 5 \text{ mph.}$$

This rate, 5 mph, is a rate of *speed*.

The same two quantities could be compared by division with the numerator and the denominator interchanged:

$$\frac{4 \text{ hours}}{20 \text{ miles}} = \frac{0.2 \text{ hours}}{1 \text{ mile}} = 0.2 \text{ hours per mile or } 0.2 \text{ hr/mi}$$

Since 0.2 hours is 12 minutes, this is equivalent to 12 minutes per mile. Runners often measure how fast they are going in minutes per mile rather than miles per hour.

In general, a rate is a quantity formed from two other quantities by division.

Example

For each rate quantity in the left column, the middle column shows the number and the right column shows the ratio of units.

Example Expressing a Rate Quantity	Number	Unit
Every student had 5 pencils.	5	$\frac{\text{pencils}}{\text{student}}$
The car's average speed on the trip was 60 miles per hour.	60	$\frac{\text{miles}}{\text{hour}}$
You paid 25 cents for each pen.	25	$\frac{\text{cents}}{\text{pen}}$

Each of these rate quantities is expressed with the special word *per*.
The term *per* means "for each."

- If each student has 5 pencils, it means there are 5 pencils for each student or 5 pencils per student.
- If the average speed was 60 miles per hour, it means the car went 60 miles for each hour of the trip or 60 miles per hour.
- If you paid 25 cents per pen, it means the cost was 25 cents for each pen bought or 25 cents per pen.

Time Rates

Rates with time in the denominator tell you how quickly things change as time passes.

Example

Rate of pay: $\frac{\$750}{20 \text{ hours}} = \frac{\$37.50}{1 \text{ hour}}$ = \$37.50 per hour

Rate of water flow: $\frac{50 \text{ gallons}}{2 \text{ hours}} = \frac{25 \text{ gallons}}{1 \text{ hour}}$ = 25 gallons per hour

Unit Costs

Unit cost is the amount of money for one unit of what is being purchased.

Example

Unit price for a pound of meat: $\frac{\$4.40}{0.5 \text{ pounds}} = \frac{\$8.80}{1 \text{ pound}}$ = \$8.80 per pound

Unit price for a can of fruit: $\frac{\$8.40}{12 \text{ cans}} = \frac{\$0.70}{1 \text{ can}}$ = \$0.70 per can or 70 cents per can

Conversion Rates

An example of a conversion rate is 12 inches per foot, the rate you use when converting a length measured in feet to the same length measured in inches.

Example

Convert 9 feet to inches:

$$9 \text{ ft} \bullet \frac{12 \text{ in}}{1 \text{ ft}} = 9 \bullet 12 \text{ in} = 108 \text{ in}$$

The following table can be used to change money between US dollars and European euros. (Note: money exchange rates often change over time.)

Dollars	$1	$5	$10	$1.25	$12.50
Euros	0.80 euros	4 euros	8 euros	1 euros	10 euros

From the table, you can see that the exchange rates are:

0.80 euros per dollar ($1.00) and $1.25 per euro

Example

Convert $200 to euros:

$$\$200 \bullet \frac{0.80 \text{ euros}}{\$1} = 200 \bullet 0.80 \text{ euros} = 160 \text{ euros}$$

Showing Rates on Graphs

This graph shows the performance of a world-class cyclist over a 1000-meter time trial. You can use the 200-meter lap times to calculate the cyclist's speed in meters per second during a 5-lap race. In this case, the cyclist maintained a constant speed throughout the trial.

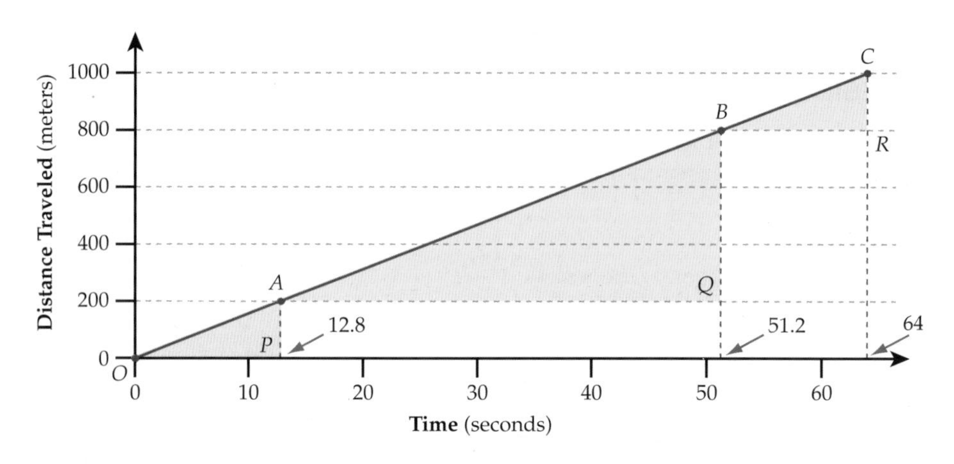

The first lap is shown on the graph by triangle *OPA*.

The speed is the ratio of $\frac{\text{distance}}{\text{time}}$ or $\frac{AP}{OP}$ on the $\triangle OPA$.

That is $\frac{200 \text{ meters}}{12.8 \text{ seconds}} = \frac{15.625 \text{ meters}}{1 \text{ second}} = 15.625$ meters per second.

Laps 2, 3, and 4 are represented on the graph by the ratio $\frac{BQ}{AQ}$ in $\triangle AQB$.

$$\text{Speed} = \frac{(800 - 200) \text{ meters}}{(51.2 - 12.8) \text{ seconds}} = \frac{600 \text{ meters}}{38.4 \text{ seconds}} = \frac{15.625 \text{ meters}}{1 \text{ second}}$$

$= 15.625$ meters per second

Lap 5, the last lap, is represented on the graph by ratio $\frac{CR}{BR}$ in $\triangle BRC$.

$$\text{Speed} = \frac{(1000 - 800) \text{ meters}}{(64.0 - 51.2) \text{ seconds}} = \frac{200 \text{ meters}}{12.8 \text{ seconds}} = \frac{15.625 \text{ meters}}{1 \text{ second}}$$

$= 15.625$ meters per second.

All of the answers are the same because the cyclist's speed in this case was constant. The graph is a straight line. The speed represents the slope of the graph. It is the same no matter which two points are used to calculate it. The straight-line graph indicates that throughout the whole race, the cyclist is traveling at a constant rate of 15.625 meters per second.

Constant Rates and Changing Rates

Water is poured into three vases of different shapes at a constant rate of 50 milliliters per second. The following pictures show the shapes of the vases. Before you look at the graphs at the right, predict how the depth of water in each vase increases with time.

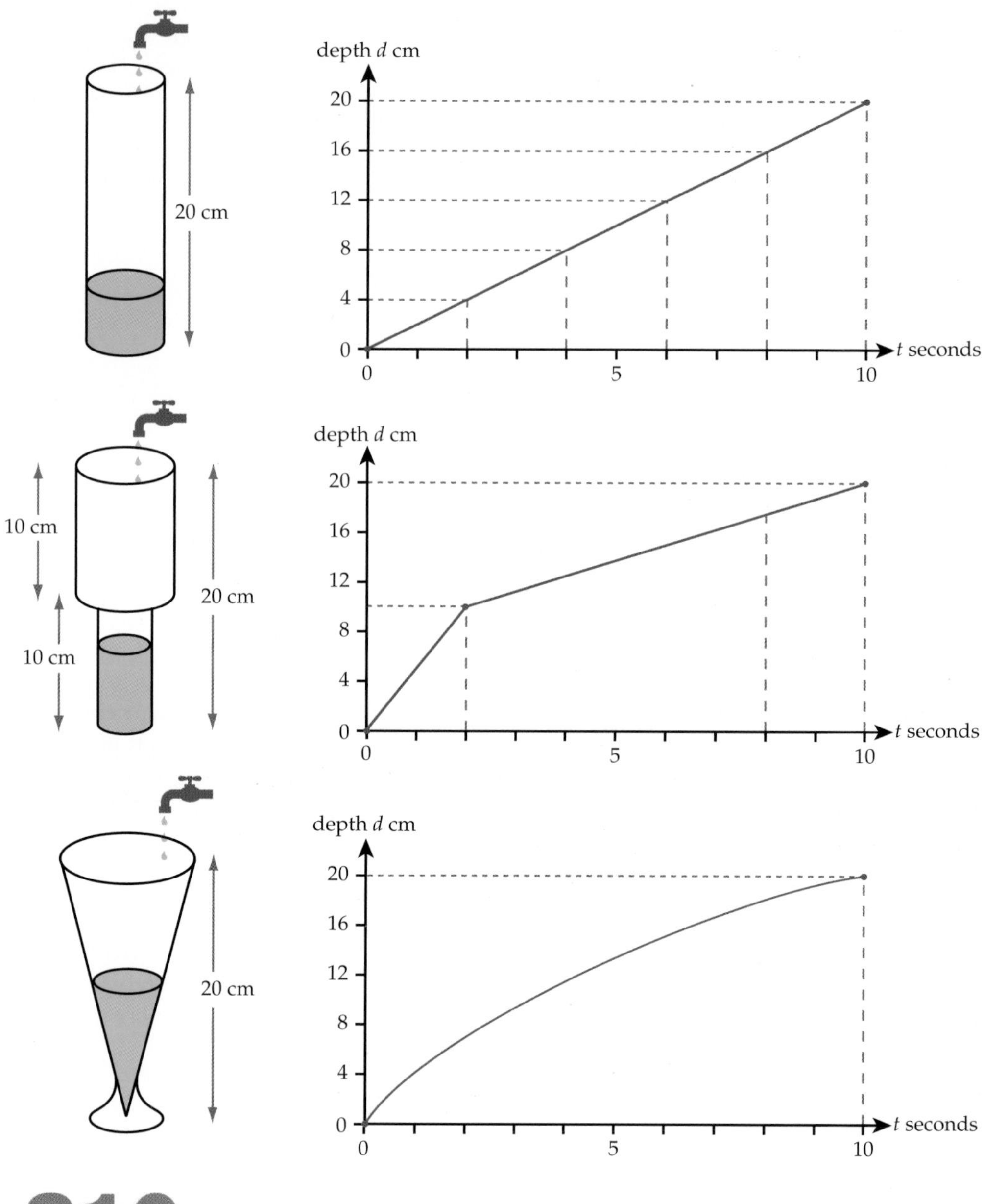
20 cm
depth d cm
20
16
12
8
4
0
t seconds
0
5
10
10 cm
10 cm
20 cm
depth d cm
20
16
12
8
4
0
t seconds
0
5
10
20 cm
depth d cm
20
16
12
8
4
0
t seconds
0
5
10

All three vases fill in 10 seconds, so the average rate of increase of depth in each vase is:

$$\frac{20 \text{ cm}}{10 \text{ sec}} = 2 \text{ cm per sec}$$

The first vase increases in depth at this rate during the whole 10 seconds.

For the second vase, the depth of water starts increasing at a rate of $\frac{10 \text{ cm}}{2 \text{ sec}} = 5$ cm per sec, but then suddenly changes to a slower rate of $\frac{10 \text{ cm}}{8 \text{ sec}} = 1.25$ cm per sec.

For the third vase, the depth of water increases quickly at first but then gradually slows down. No part of the graph is straight, so the rate of increase in depth is never constant.

The Algebra of Constant Rates

The $y = mx$ Family of Functions

The function $i = 12f$ represents the relationship between two variables, i and f, with 12 as a constant.

- Variable i is a length measured in inches.
- Variable f is the same length measured in feet.
- The 12 is the conversion rate, a constant with the unit *inches per foot*.

This function is one of a family of functions that are in the form $y = mx$, where:

- x and y are the variables.
- m is a constant, the constant rate.
- The unit for m is "unit of y per unit of x."

Some other members of this $y = mx$ family are:

1. $d = 5t$, where d is the number of miles traveled and t is the number of hours. The constant rate of 5 is the constant speed of 5 miles per hour.

2. $C = 8.80w$, where C is the cost of meat in dollars and w is its weight, in pounds. The constant rate of 8.80 is the unit cost in dollars per pound.

3. The function $D = \frac{5}{4}E$ and its inverse $E = \frac{4}{5}D$ could both represent the relationship between US dollars and European euros. In these functions, the constant rates of exchange are $\frac{5}{4}$ dollars per euro and $\frac{4}{5}$ euros per dollar.

The $y = mx + n$ Family of Functions

The following table is the relationship between temperatures measured on the Fahrenheit scale and temperatures measured on the Celsius scale.

	Water Freezes							**Water Boils**
Temperature in ° C	0	5	10	15	20	25	30	100
Temperature in ° F	32	41	50	59	68	77	86	212

In the table, the constant rate of change is an increase of 9 degrees in the value of F for every increase of 5 degrees in the value of C.

The function for this relationship is $F = \frac{9}{5}C + 32$, where:

- The constant rate of change is $\frac{9}{5}$ degrees F per degree C.
- The other constant, 32, is required because the zero on the Fahrenheit scale is not set at the freezing point of water.

This function is one of a family of functions that take the form $y = mx + n$, where:

- x and y are the variables.
- m is a constant, the constant rate of change.
- The unit for m is "unit of y per unit of x."
- The n is another constant.

Another example of this $y = mx + n$ family of functions is $d = 5t + 2$, where:

- d is the number of miles traveled since the clock started.
- t is the number of hours since the clock started.
- The constant rate of 5 is the constant speed of 5 miles per hour.
- 2 miles had already been traveled when the clock started at $t = 0$.

PROPORTIONAL RELATIONSHIPS

CHAPTER 19

In Chapter 17, *Ratio*, you learned that a ratio is the comparison of two quantities by division. This chapter will focus on ratios that are equal, unchanging, or "constant."

When the ratio between two varying quantities is constant, the relationship is called a *proportional relationship*.

Example

A store has 22 tricycles on display. How many wheels are in that display?

There will be 3 times as many wheels as tricycles because each tricycle has 3 wheels. 22 tricycles will have 66 wheels.

The ratio of wheels to tricycles is constant, regardless of how many tricycles there are. Thus, if you know the number of tricycles, you can multiply this number by 3 to find the number of wheels.

Similarly, if you know the number of wheels, then you can divide this number by 3 to find the number of tricycles.

A relationship between two quantities, like the number of tricycles and the number of wheels, can be expressed by a constant ratio, wheels to tricycles = 3 : 1.

This is an example of a proportional relationship. You say, "The number of wheels is proportional to the number of tricycles."

You can also express the relationship as the ratio of tricycles to wheels: 1 to 3, or 1 : 3. You say, "The number of tricycles is proportional to the number of wheels."

Chapter 19

Representing Proportional Relationships in Ratio Tables

A ratio table shows the equal ratios in a proportional relationship. Here is a ratio table for the tricycles-to-wheels relationship.

Tricycles	0	1	10	11	22	n
Wheels	0	3	30	33	66	$3n$

Points to note about ratio tables:

- Every ratio table has (0, 0) as a pair of corresponding values.
- Each column in a ratio table has two numbers. The number in the bottom row is always the result of multiplying the number in the top row by some constant. In the table above, this constant is 3.
- You can find new pairs by multiplying any existing pair of values (except 0) by the same number.
- You can also find new pairs by adding (or subtracting) other pairs in the table.
- You can use these strategies to find any unknown quantity, such as 66 wheels for 22 tricycles in the table above.

Representing Proportional Relationships with Formulas

The proportional relationship between tricycles, (t), and wheels, (w), can be represented using either of two alternative formulas:

$\frac{w}{t} = 3$ or $w = 3t$ — Here, the constant of proportionality is 3.

$\frac{t}{w} = \frac{1}{3}$ *or* $t = \frac{1}{3}w$ — Here, the constant of proportionality is $\frac{1}{3}$. Note that $\frac{1}{3}$ is the reciprocal of 3.

Either one of these formulas can be used to find an unknown quantity.

Example

Put $w = 78$ into the formula $t = \frac{w}{3}$: $t = \frac{78}{3}$

Simplify: $t = 26$

Solution: For 78 wheels there will be 26 tricycles.

Generalizing from the example:

If quantity q is proportional to quantity p, then the two quantities have a constant ratio.

Constant ratio: $\frac{p}{q} = k_1$

The ratio of p to q is always k_1.

This also means that one quantity is a constant multiple of the other.

Constant multiple: $p = k_1 q$ p is always a multiple of q.

Similarly, quantity p will be proportional to quantity q.

Constant ratio: $\frac{q}{p} = k_2 = \frac{1}{k_1}$

The ratio of q to p is always k_2, where $k_2 = \frac{1}{k_1}$.

Constant multiple: $q = k_2 p$

q is always a *multiple* of p, where $k_2 = \frac{1}{k_1}$.

Representing Proportional Relationships with Graphs

The graph for any proportional relationship between two quantities is a straight line that starts at the origin, (0, 0).

Example

The following graphs show the proportional relationship between tricycles and wheels.

The graph for $w = 3t$ has t on the x axis and w on the y axis.

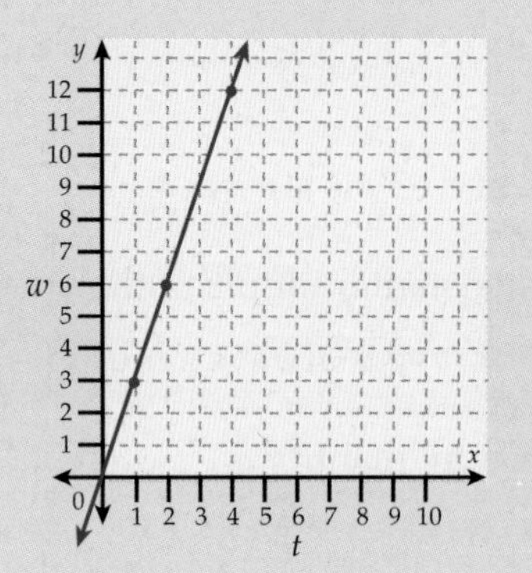

The graph for $t = \frac{1}{3}w$ has w on the x axis and t on the y axis.

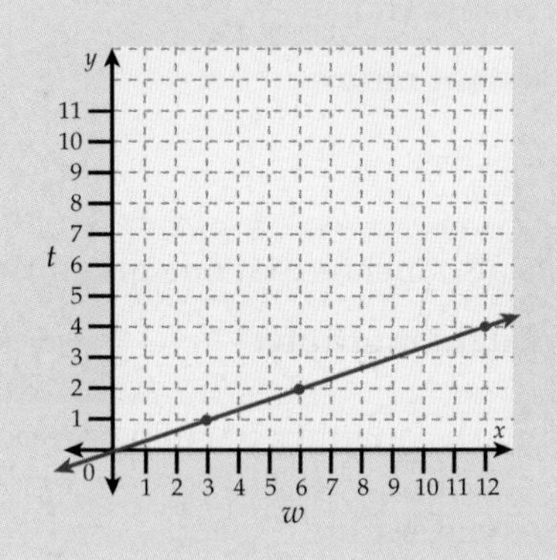

Each column in the earlier table corresponds to one of the points on the graph.

In the left graph, there is a point at (1, 3), one at (2, 6), one at (4, 12), and so on.
In the right graph, there is a point at (3, 1), one at (6, 2), one at (12, 4), and so on.

Look at the *slopes* of the graphs, where slope = $\frac{\text{rise (vertical)}}{\text{run (horizontal)}}$ between any two points on the graph. (See Chapter 20, *Graphing Relationships*.)

In both graphs, the slopes are constant, since the lines are straight.

In the graph on the left, the slope is 3, and the line passes through (0, 0), so 3 is the constant of proportionality.

In the graph on the right, the slope is $\frac{1}{3}$, and the line passes through (0, 0), so $\frac{1}{3}$ is the constant of proportionality.

In general, if quantity p is proportional to quantity q with formula $p = kq$, then the slope of the graph will be equal to k, the constant of proportionality.

Graphs can be used to find approximate values of unknown quantities in proportional relationships. For example, reading either graph at $w = 20$ would tell you that for 20 wheels you would have less than 7 tricycles. Since tricycles must have three wheels, you would only have 6 complete tricycles.

Common Situations with Proportional Relationships

Many problems describe situations where one quantity is proportional to another. The situations may seem very different, but they all use formulas like $p = kq$ as described above, although the letters used for the quantities may be different, and the constant of proportionality, k, may be replaced by a given number.

It is helpful to group these relationships into two types one where the quantities are of the *same* dimension, the other where the quantities are of *different* dimensions. (For more information on dimensions, see Chapter 16, *Units and Quantities*.)

Chapter 19

Proportional Relationships between Quantities of the Same Dimension

In these situations, the constant of proportionality, k, is dimensionless: $p = kq$.

1. The quantities p and q are both counts of things:
   a. Tables/chairs: k is the number of chairs at each table. Different types of tables have different k's, but $k = 4$ for a common type of table.
   b. Cars/wheels: k is the number of wheels on a car. Different types of cars have different k's, but $k = 4$ for most cars.
   c. Students/teachers: k is the student/teacher ratio.
2. The quantities p and q are both lengths, and k expresses a size relationship, called the *scale factor* or *similarity ratio*:
   a. Similar triangles: Corresponding sides are proportional. k is the ratio of similarity.
   b. Maps or floor plans: k is the scale factor.
   c. Enlargement/reduction: k is a photocopier setting (often given as a percent).
3. The quantities p and q are both lengths, and k expresses a shape relationship:
   a. Steepness of ramps, stairs, roads, or roofs: k is the slope, and is expressed in at least two different ways (a 6% grade on a road; a 1 to 5 slope of a roof).
   b. Slope: In a simple case, where vertical and horizontal scales are the same, the mathematical slope of the graph is the same as the geometric slope. However, you have to be careful because, if vertical and horizontal scales are not the same, the mathematical slope is different from the geometric slope.
4. The quantities p and q are both monetary values:
   a. Cost/sales tax: k is the sales tax rate, usually expressed as a percent.
   b. Cost/discount: k is the percent taken off in a sale ("every item 20% off").

5. The quantities p and q are any quantities of the same type, and k is the *conversion factor*.

   a. cm/inch: $c = 2.54i$, where i is a length measured in inches and c is the same length measured in centimeters. Notice that the value of c will always be 2.54 times the corresponding value of i. This is because centimeters are shorter than inches. A stick that is 2 inches long is 5.08 cm long.

   b. min/hour: $m = 60h$, where h is a time interval in hours and m is the same time interval in minutes. There are always 60 minutes in an hour.

Proportional Relationships between Quantities of Different **Dimensions**

In these situations, the constant of proportionality, k, has units expressed with the special word "per." That is, if

$$p = kq$$

then the constant of proportionality, k, gives the amount of p per unit amount of q.

$$k = \frac{p}{q}$$

These are what are usually called rates. A rate is a "per unit" quantity: an amount of one quantity per unit amount of another quantity. (See Chapter 18, *Rates*. Also see Chapter 16, *Units and Quantities* for explanations of calculating with units.)

Identifying Proportional Relationships

Being able to determine whether or not a proportional relationship exists between two varying quantities is an important aspect of mathematics.

There are three ways to do this:

- If the graph of the relationship between the quantities is a line that passes through the point (0, 0) then you can be sure that the relationship between the graphed quantities is proportional.

- If you can express one quantity in terms of another using a formula of the form $y = kx$, then you can be sure that the relationship between the quantities is proportional.
- If you find that the ratios between the varying quantities is constant, you can be sure that the relationship is proportional.

If a line represents a relationship between two varying quantities, the relationship is proportional if, and only if, the line passes through the point (0, 0). Note: Sometimes, for emphasis, a proportional relationship is called *directly proportional.*

Non-proportional Relationships

When the ratios between varying quantities are not equal to a constant, you can have a variety of different types of *non-proportional relationships.*

Area and Radius of a Circle

Perhaps the clearest way to consider the relationship between these two varying quantities is to examine their graph.

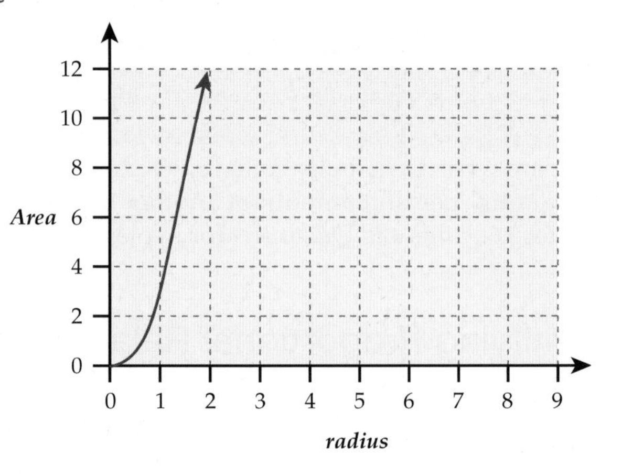

You can see that the graph passes through the point (0, 0) but it is a curve, not a straight line.

This graph shows that the relationship between the area and the radius of a circle is not a directly proportional relationship. The relationship between the area and the radius of a circle is a *quadratic* relationship.

Linear Relationships that Are Not Directly Proportional Relationships

The type of non-proportional relationships that *are* represented by lines is an important type of relationship. Such lines, however, do not pass through (0,0).

Example

Suppose you go bowling, and it costs you $2 to rent shoes and $4.50 to play each game.

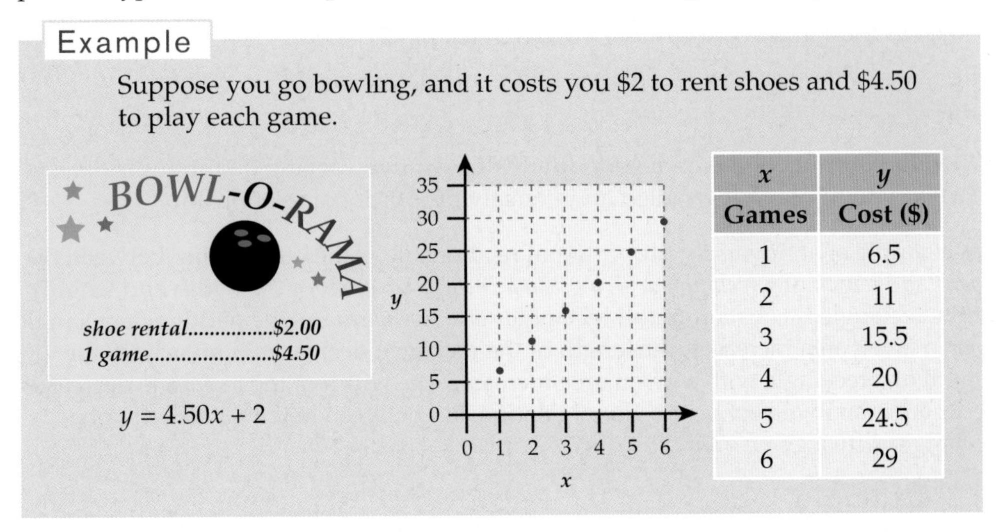

x	*y*
Games	**Cost ($)**
1	6.5
2	11
3	15.5
4	20
5	24.5
6	29

In the graph, the data lie along a line that does not pass through (0, 0). The ratio between amount spent and the number of games is not equal to a constant. Thus, the relationship between amount spent and the number of games is not directly proportional.

Situations such as the bowling scenario are modeled by the general formula: $y = kx + b$. This form is closely related to $y = kx$, the formula that represents a directly proportional relationship. Nonetheless, $y = kx + b$ consists of a proportional part, kx, and a constant part, b. When this constant part b is not equal to zero, it disrupts the relationship and makes y and x not directly proportional.

You may have realized from looking at the graph that the constant, b, is the *y- intercept*. When the value of b is positive, the line cuts the y-axis above (0, 0). When the value of b is negative, the line cuts the y-axis below (0, 0).

If a line represents a relationship between two varying quantities, the relationship is directly proportional, *if and only if*, the line passes through the point (0, 0). If the line does not pass through (0,0), then the relationship between the quantities is not directly proportional.

Inversely Proportional Relationships

Inversely proportional relationships are a class of relationships that are not directly proportional.

When the product of the corresponding values of two varying quantities is *equal to a constant* you have what is called an inversely proportional relationship.

An example of an inversely proportional relationship is the relationship between the *width* and *length* of a *rectangle with a given area*. The product of the width and length of the rectangle is constant for all rectangles *of a given area*. As the width of a rectangle with a given area increases, the length of the rectangle decreases. Similarly, as the length of a rectangle with a given area increases, the width of the rectangle decreases. So, there is an inversely proportional relationship between length and width of a rectangle with a given area.

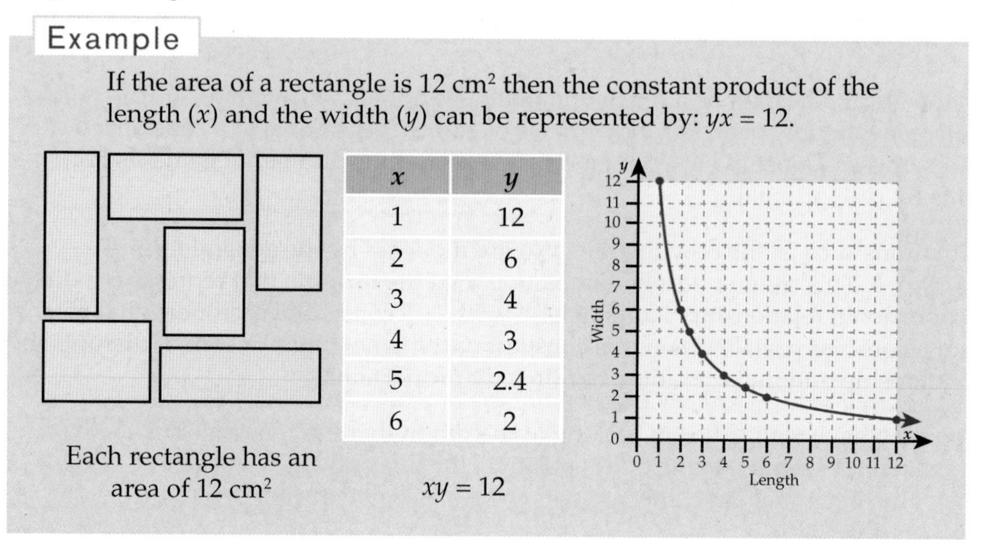

x	y
1	12
2	6
3	4
4	3
5	2.4
6	2

$xy = 12$

Any situation involving a constant product can also be described in terms of a *constant multiple*. Another way of saying that the product of the length and width of a rectangle with fixed area is equal to a constant is to say that the length (x) is a constant multiple of the *inverse* (or *reciprocal*) *of the width* (y). In terms of rectangles with a fixed area of 12 cm^2, this fact can be represented with the formula: $x = \frac{12}{y}$.

If one quantity varies inversely with the other quantity, then one quantity is a *rational function* of the other quantity. Any two such quantities, x and y, are represented by the general formula:

$$y = \frac{k}{x}$$

where k is a fixed number not equal to 0.

In this formula, x and y are variables and k is the *constant of proportionality*. In some textbooks, k is called the *constant of variation*.

Note that $y = \frac{k}{x}$ can also be written as $xy = k$.

The graph of quantities that vary inversely is called a *hyperbola*.

A graph that is a hyperbola comes very close to, but does not intersect, the vertical or horizontal axis because the width or length of a rectangle with area 12 cm^2 can never equal 0. In this case, the vertical and horizontal axes are called the *asymptotes* of the hyperbola.

CHAPTER 20

GRAPHING RELATIONSHIPS

What Is Slope?

Slope is a measure of steepness.

Slope is the relationship between how far *up* something rises and how far *along* it runs. It is the ratio of *rise* (change in the vertical direction) to *run* (change in the horizontal direction).

Example

These are examples of things that have steepness:

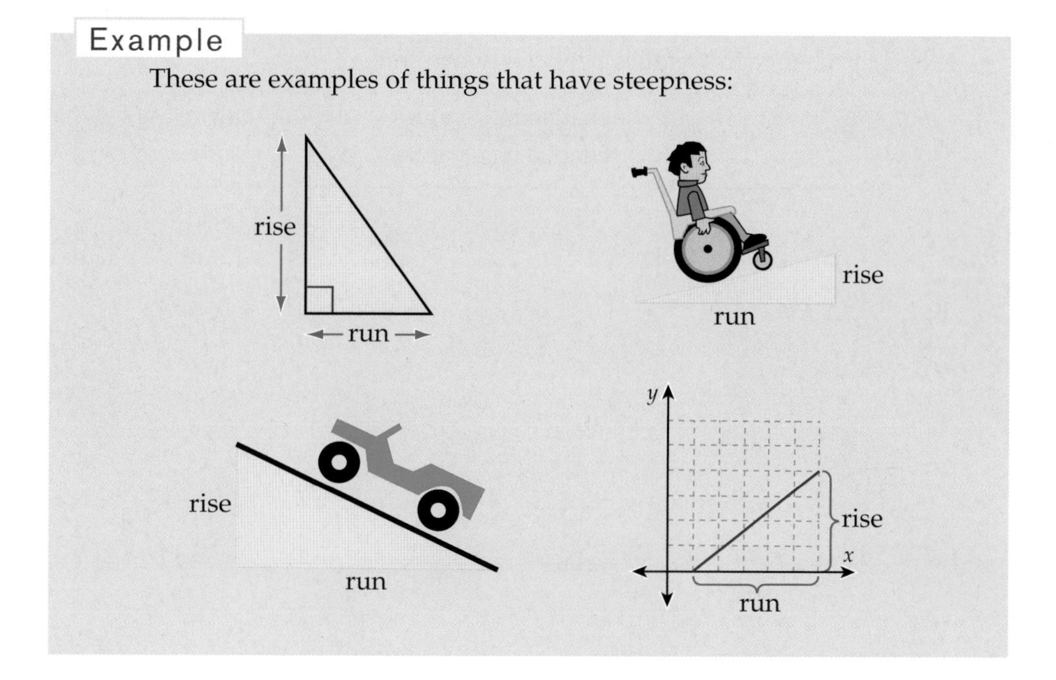

Slope is a ratio.

The comparison of rise and run determines the slope.

$$slope = \frac{\text{rise}}{\text{run}}$$

Slope is the ratio of vertical change (*rise*, shown on the y-axis) to horizontal change (*run*, shown on the x-axis).

The slope of a line is the same between any two points on the line.

For any two points (x_1, y_1) and (x_2, y_2) on a line:

- Rise is the change in y-values between two points: $y_2 - y_1$.
- Run is the change in x-values between two points: $x_2 - x_1$.

$$slope = \frac{\text{rise}}{\text{run}} = \frac{\text{change in } y\text{-values}}{\text{change in } x\text{-values}} = \frac{y_2 - y_1}{x_2 - x_1}$$

Slope and Similar Triangles

A good way to understand slope geometrically is through similar triangles.

Given a straight line in the coordinate plane, every right triangle with a hypotenuse that extends between two points on that line is similar to any other such triangle. Since these triangles are similar, corresponding sides have the same ratio. This means that the ratio of rise to run will be equal for all such triangles on that line.

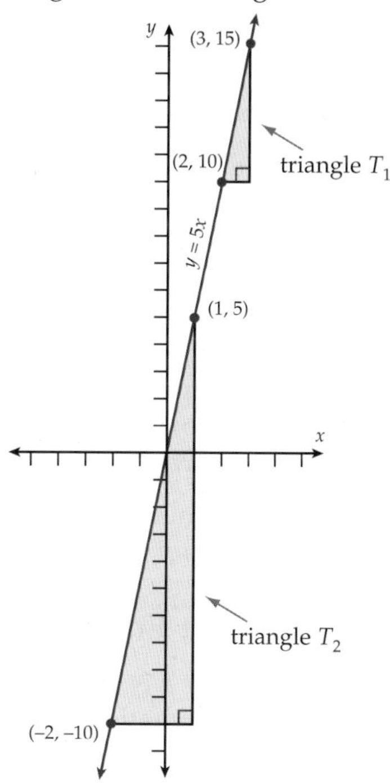

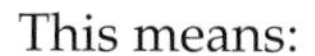

This means:

- The hypotenuse of every such triangle has the same slope.
- The slope between any two points on a straight line is the same as between any other two points.

Steepness of a Line

There are at least four ways to represent the steepness of a line:

1. The slope expressed as a fraction: for example, $\frac{2}{5}$
2. The slope expressed as a decimal: for example, 0.4
3. The slope expressed as a percent, or *percent grade*: for example, 40%
4. The angle of elevation: for example, approximately 22°

The fourth way is different from the other three. The angle of elevation is not called a slope.

There are several ways to calculate the steepness of a line from the graph of the line on a coordinate plane:

Formula

1. Choose any two points on the line, (x_1, y_1) and (x_2, y_2).
2. Then calculate slope = $\frac{y_2 - y_1}{x_2 - x_1}$

Right Triangle with Base Length of One Unit

1. Choose one point on the line.
2. From your point, draw a line segment that extends one unit to the right (one step horizontally) and from there, draw a line segment that extends up or down until you reach the line. You will have drawn a right triangle with a hypotenuse that extends along the line.
3. Count the number of units you moved up or down to reach the line. That number is the slope of the line.

Slope-Intercept Equation

$y = mx + b$ is the *slope-intercept equation* of the line.
In this equation, the coefficient of x is the slope of the line: m = slope.

Angle of Elevation

With a protractor, measure the angle made between the line and the x-axis. This *angle of elevation* is an alternative measure of steepness. It is not called a slope.

The Slope-Intercept Equation of a Line

Every line in the coordinate plane can be defined by an equation.

An equation is a statement of the relationship between x values and y values.

Example

$y = 2x + 1$

This equation states that y is equal to 1 more than 2 times x.
The slope is 2. The y-intercept is 1.

In this equation, if $x = -3$, then $y = 2(-3) + 1 = -6 + 1 = -5$. The ordered pair $(-3, -5)$ satisfies this equation. The point with these coordinates lies on the line.

An (x, y)-table shows additional ordered pairs of x-values and y-values that satisfy the equation.

x	–3	–2	–1	0	1	2	3
y	–5	–3	–1	1	3	5	7

Each pair of (x, y)-values in this (x, y)-table gives the x-coordinate and y-coordinate of a point that lies on the line defined by the equation $y = 2x + 1$.

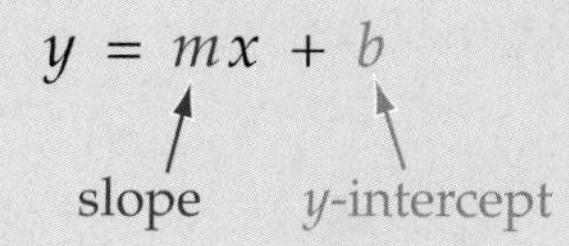

The *slope-intercept equation* of a line has the form: $y = mx + b$.

The y-intercept of a line is the y-coordinate where the line crosses the y-axis.

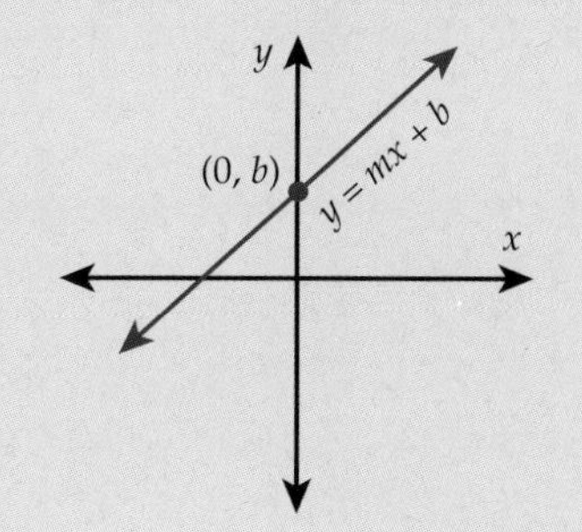

Summary of $y = 2x + 1$

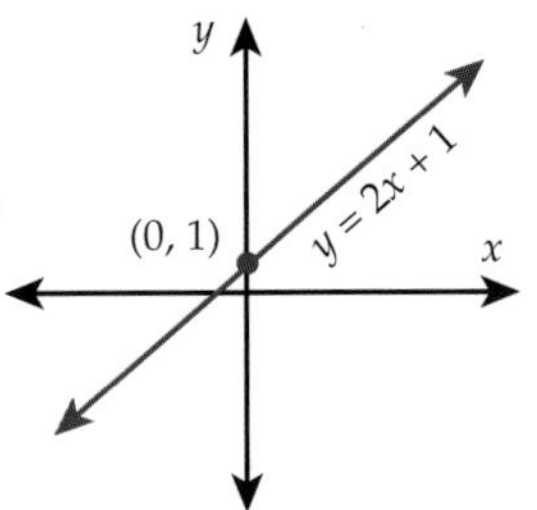

In the equation $y = 2x + 1$, the slope is 2 and the y-intercept is 1. As shown in the (x, y)-table for this equation, if $x = 0$, then $y = 1$. The graph of the line crosses the y-axis at the point (0, 1).

Equations in the Form $y = mx + b$

You can easily identify the slope and y-intercept of a line if the equation of the line is in the form, $y = mx + b$.

Example

- Suppose the equation is $y = x - 1$. The slope is 1, and the y-intercept is –1.
- Suppose the equation is $y = -2x + 3$. The slope is –2, and the y-intercept is 3.
- Suppose the equation is $y = \frac{3}{5}x$. The slope is $\frac{3}{5}$, and the y-intercept is 0.

An equation of a line states the relationship between the (x, y)-coordinates of every point on the line.

The *Y*-Intercept

In an (x, y)-table, the y-intercept is the y-value when $x = 0$.

Example

In the table showing x-values and y-values that satisfy the equation $y = 2x + 1$, you can see that $y = 1$ when $x = 0$. The y-intercept is 1.

x	–3	–2	–1	**0**	1	2	3
y	–5	–3	–1	**1**	3	5	7

When $x = 0$ for any point (x, y), that point is on the y-axis.

In other words, every point on the y-axis has an x-coordinate of 0.

Notice that every point with an x-coordinate of 0 is on the y-axis. This means that $x = 0$ for every point (x, y) that is on the y-axis.

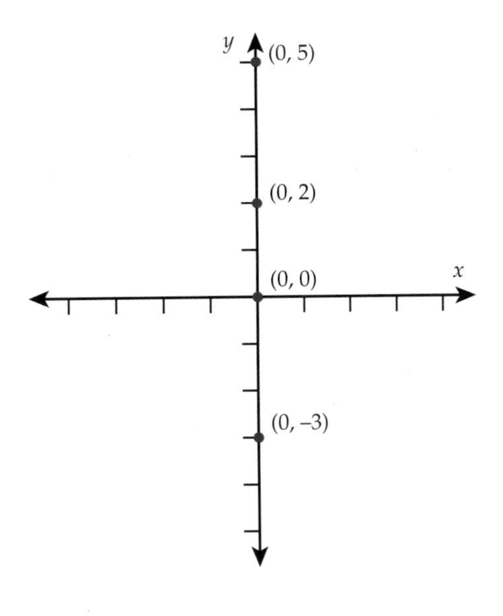

REPRESENTING DATA

CHAPTER 21

Organizing Data

American Lance Armstrong is famous for winning the international bicycle race, the Tour de France, many times. You can marvel at Armstrong's physical achievements. You cannot, however, say that everyone in the United States is a great cyclist because of Armstrong's many victories.

The achievements of a single person cannot be generalized to reflect the achievements of a whole group.

When looking closely at the characteristics of a group, with the goal of saying something about the entire group, you need techniques for examining the distribution of the data and for describing trends in data. The range, mean, median, and mode can help you interpret and understand the characteristics of data.

Raw Data

This table gives the populations of the fifty states of the United States and the District of Columbia.

Table 1. Alphabetical List of States, with Population in Millions

State / District	Pop.	State	Pop.	State	Pop.
Alabama	4.45	Kentucky	4.04	North Dakota	0.64
Alaska	0.63	Louisiana	4.47	Ohio	11.35
Arizona	5.13	Maine	1.27	Oklahoma	3.45
Arkansas	2.67	Maryland	5.30	Oregon	3.42
California	33.87	Massachusetts	6.35	Pennsylvania	12.28
Colorado	4.30	Michigan	9.94	Rhode Island	1.05
Connecticut	3.41	Minnesota	4.92	South Carolina	4.01
Delaware	0.78	Mississippi	2.84	South Dakota	0.75
Dist. of Columbia	0.57	Missouri	5.60	Tennessee	5.69
Florida	15.98	Montana	0.90	Texas	20.85
Georgia	8.19	Nebraska	1.71	Utah	2.23
Hawaii	1.21	Nevada	2.00	Vermont	0.61
Idaho	1.29	New Hampshire	1.24	Virginia	7.08
Illinois	12.42	New Jersey	8.41	Washington	5.89
Indiana	6.08	New Mexico	1.82	West Virginia	1.81
Iowa	2.93	New York	18.98	Wisconsin	5.36
Kansas	2.69	North Carolina	8.05	Wyoming	0.49

This table gives each state's population in millions, at the time of the U.S. Population Census on April 1, 2000. You can see that the population of Arizona was 5.13 million and that of Ohio was 11.35 million. At this time, the population of the entire U.S. was approximately 281.4 million.

This particular way of representing the data does not allow you to say what characteristics the group possesses as a whole, or allow you to see what relationships there might be between the populations of the different states. In order to be able to interpret this set of data, you must organize it.

The following table shows the same data, but with the states ordered by increasing population:

Table 2. List of States Ordered by Size, with Population in Millions

State / District	Pop.	State	Pop.	State	Pop.
Wyoming	0.49	Utah	2.23	Missouri	5.60
Dist. of Columbia	0.57	Arkansas	2.67	Tennessee	5.69
Vermont	0.61	Kansas	2.69	Washington	5.89
Alaska	0.63	Mississippi	2.84	Indiana	6.08
North Dakota	0.64	Iowa	2.93	Massachusetts	6.35
South Dakota	0.75	Connecticut	3.41	Virginia	7.08
Delaware	0.78	Oregon	3.42	North Carolina	8.05
Montana	0.90	Oklahoma	3.45	Georgia	8.19
Rhode Island	1.05	South Carolina	4.01	New Jersey	8.41
Hawaii	1.21	Kentucky	4.04	Michigan	9.94
New Hampshire	1.24	Colorado	4.30	Ohio	11.35
Maine	1.27	Alabama	4.45	Pennsylvania	12.28
Idaho	1.29	Louisiana	4.47	Illinois	12.42
Nebraska	1.71	Minnesota	4.92	Florida	15.98
West Virginia	1.81	Arizona	5.13	New York	18.98
New Mexico	1.82	Maryland	5.30	Texas	20.85
Nevada	2.00	Wisconsin	5.36	California	33.87

Frequency Distribution Tables

The information in the state populations table can be organized by sorting the populations into intervals. This table which divides data into several classes and records the frequency for each class, is called a *frequency distribution table.*

Population (millions)	Number of States
0.00 – 1.99	16
2.00 – 3.99	9
4.00 – 5.99	12
6.00 – 7.99	3
8.00 – 9.99	4
> 10.00	7
Total:	51

The range of each division is the *class interval.* This table has a class interval of 2.00 million.

The number of items in each class is the *class frequency*. The class frequency in this frequency table is the number of states that falls into each interval.

Example

The interval from 4.00–5.99 includes all of the states who have a population greater than or equal to 4.00 million and less than or equal to 5.99 million.

The class interval is 2.00 million and the class frequency for states with populations of between 4.00 million and 5.99 million is 12.

You know from the state population table that the population of Kentucky is 4.04 million. Kentucky would lie in the population interval 4.00 – 5.99 million.

There is no rule for dividing a set of data into classes or how many classes to create for any particular set of data. The number of classes that work well for any set of data depends on the set of data that is to be organized. It also depends on how the set of data is to be used. It is, however, difficult to make sense of a frequency distribution table if the class intervals are too small or too large.

Class intervals of 10 million would have been too large to produce a useful picture of state population data. If you had used class intervals of 10 million, the data would have been organized into only 4 intervals and the first interval would contain 44 of the states.

Class intervals of 1 million would have been too small to produce a useful picture of this data. If you had used class intervals of 1 million, the data would have been organized into 34 intervals, and many of these intervals would have contained no states.

Range

The numerical difference between the maximum and the minimum values included in the data is called the *range* of the distribution:

range = the maximum value – the minimum value

Example

The state with the largest population is California with 33.87 million, and the smallest is Wyoming with 0.49 million.

33.87 – 0.49 = 33.38 million

So the range is 33.38 million.

The range is sometimes expressed as an interval that shows the minimum through the maximum. In the population data, the range is:

0.49 million to 33.87 million

Histograms

A frequency distribution table can be represented as a bar graph.

This graph represents the distribution of state populations from the frequency distribution table.

This type of graph is called a *histogram*.

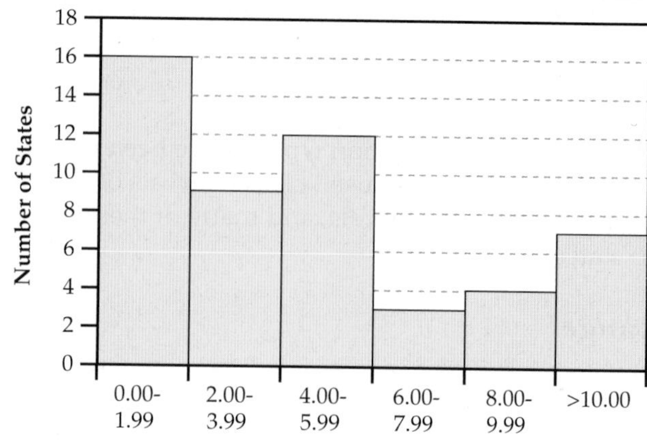

A histogram represents the frequency of each class in a diagram. The area of each rectangle is proportional to the frequency of its class.

The broken line in the second diagram connects the midpoints of the top sides of rectangles in the histogram above.

This type of broken line is called a *frequency polygon*.

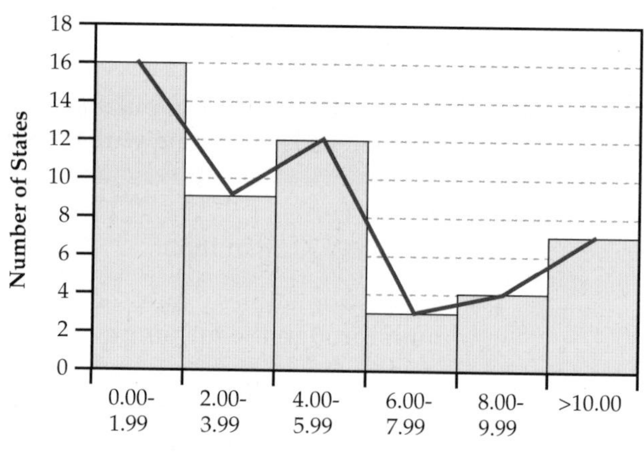

In looking at the shape of the histogram and frequency polygon, keep in mind that you combined all of the states with populations of at least 10 million into a single class. The right-hand tail (representing a small number of very high-population states) would appear much longer if you not shortened the tail by combining the last few observations into a single class.

Relative Frequency

This is the same frequency distribution table of state populations that you looked at before.

Population (millions)	Number of States
0.00 – 1.99	16
2.00 – 3.99	9
4.00 – 5.99	12
6.00 – 7.99	3
8.00 – 9.99	4
> 10.00	7
Total:	**51**

In some instances it is useful to examine the fractions of the total number of observations that fall into each class. These fractions are called the *relative frequencies*. You can use the following formula to calculate them:

$$\text{Relative frequency} = \frac{\text{frequency of class}}{\text{total number of results}}$$

The second table shows the relative frequencies for the state population table. This is called a *relative frequency distribution table.*

Population (millions)	Relative Frequency Proportion	Relative Frequency Percent
0.00–1.99	0.314	31.4%
2.00–3.99	0.176	17.6%
4.00–5.99	0.235	23.5%
6.00–7.99	0.059	5.9%
8.00–9.99	0.078	7.8%
> 10.00	0.137	13.7%
Total:	**1.000**	**100.0%**

Relative frequency distribution tables are a useful way of adjusting for differences between the numbers of observations for two different data sets. For example, if you wanted to compare the distribution of state populations to the distribution of populations by county within any one state, it would be helpful to use relative frequency distributions to make the comparison.

Cumulative Frequency

Consider how many of the states have populations that are below a certain amount.

You can extract the following information from the frequency distribution table on the preceding page:

- Population less than 2 million: 16 states
- Population less than 4 million: 16 + 9 = 25 states
- Population less than 6 million: 16 + 9 + 12 = 37 states
- Population less than 8 million: 16 + 9 + 12 + 3 = 40 states

The results of this method of organizing the data are given in the table below.

Population (millions)	Number of States
0.00–1.99	16
0.00–3.99	25
0.00–5.99	37
0.00–7.99	40
0.00–9.99	44
0.00–39.99	51

When you add the values of the individual classes in a frequency table so a column reflects the total value of the all the columns preceding it, the table is called the *cumulative frequency* table.

Mean

You often compare frequency distributions based on their means.

The *mean* is defined as the sum of all of the observations divided by the number of observations.

Example

The total population of the U.S. at the time of the 2000 census was 281.4 million and there are 51 observations. (That is, fifty states and the District of Columbia.)

So the mean of this distribution is:

$$\text{mean} = \frac{281.4}{51} = 5.52 \text{ million}$$

The *mean* of a distribution is also sometimes called the *average*.

Estimating the Mean from a Frequency Distribution Table

You can *estimate the mean* from a frequency distribution table that summarizes the data without knowing all of the individual observations. You can do this by choosing a middle value for each class and using it to represent the value of all the data in that particular class. For example, you could use the following approximate representation for the data in the frequency distribution table for state populations.

Here, you have chosen the midpoint for the first five class intervals. For the last interval, the states with populations of 10 million and more, there is no obvious *middle value*. You choose 18 million as a representative middle value for this group by choosing a value nearly in the middle of the seven observations.

Population (millions)	**Middle Value** (millions)	**Number of States**	**Estimated Class Population** (millions)
0.00–1.99	1.0	16	16
2.00–3.99	3.0	9	27
4.00–5.99	5.0	12	60
6.00–7.99	7.0	3	21
8.00–9.99	9.0	4	36
> 10.00	18.0	7	126
Total:		**51**	**286**

The estimated total U.S. population calculated using the frequency table and approximate middle values shown above is 286 million. The estimated mean population is 5.6 million. Interestingly, these results come very close to matching the actual total of 284.1 million and the actual mean of 5.52 million.

You can *estimate the mean*, given only the summary information of a frequency distribution table, by carrying out the following steps:

Step 1. For each class, pick a middle value that reasonably represents the entire class.

Step 2. For each class, multiply the value chosen in Step 1 by the class frequency, and add the products to give an estimated sum for the entire population.

Step 3. Divide the result found in Step 2 by the total number of observations to estimate the mean of all of the observations.

When you take one numerical value, such as the mean, to represent the entire body of data to be examined, it is called a representative value or statistic. The median and mode are two other examples of representative values that can be used to characterize given data.

Median

When a set of data is arranged by numerical value, the value that falls in the middle of the data is called the *median*.

Example

To find the median for the U.S. state populations:

- Arrange the states in order by the population size.
- Find the 26th observation.
  In this case, South Carolina is the 26th observation and is the U.S. state with the median population.

When you have an even number of values, the median is the mean of the two values in the middle.

A cumulative frequency distribution table allows the median to be located easily. Refer to the cumulative frequency distribution table for U.S. state populations. Notice that there are exactly 25 states with populations of less than 4 million. Counting the District of Columbia, the 26th state (the first one with a population of 4 million or more) is the U.S. state with the median population. This turns out to be South Carolina, with a population of 4.01 million at the time of the 2000 census. The median state population is 4.01 million.

Population (millions)	**Number of States**
0.00 – 1.99	16
0.00 – 3.99	25
0.00 – 5.99	37
0.00 – 7.99	40
0.00 – 9.99	44
0.00 – 39.99	**51**

Mode

In a set of data, the value that occurs most often is called the *mode*.
In a frequency distribution table, the value interval with the greatest frequency is called the *modal class*.

Example

In U.S. state populations, the modal class is the very first interval representing the relatively large number of states with populations of less than 2 million.

The mode is the value for which the frequency histogram reaches its highest point, 1.00 million. Notice that nearly three-quarters of the states (37 of 51) have populations of less than 6 million and two-thirds (34 of 51) have populations that are smaller than the mean, 5.52 million.

The distribution of state populations is relatively one-sided, with the largest number of observations clustered at one end of the distribution.

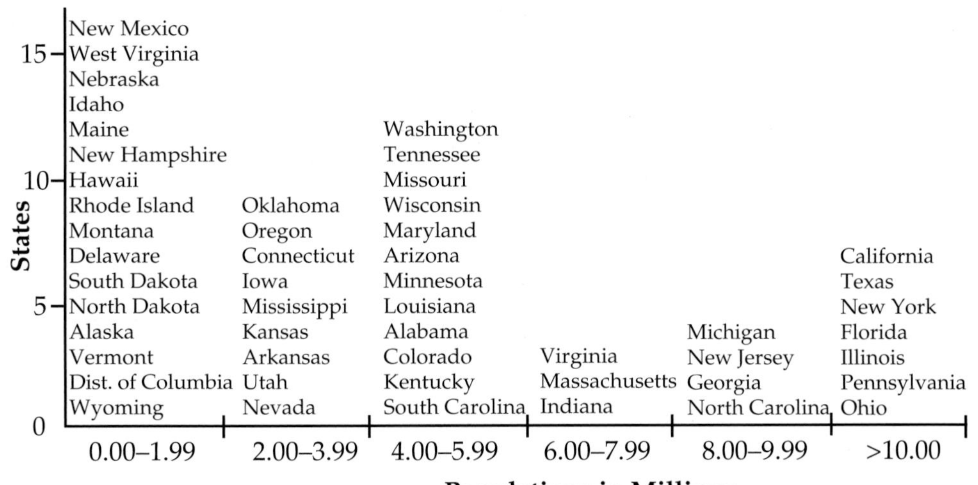

Other data sets are likely to be more centrally distributed than the U.S. population data.

Example

The heights of all students in one class are likely to be centrally distributed.

The mode is less useful than the mean and median for comparing data sets for these reasons:

- Some data sets do not have one single value that occurs most often. For example, a *bi-modal* data set has two values with equal highest frequency.
- Some sets of measurement data do not have a mode at all.

Example

Below are some data that are the weights (in kilograms) of twelve newborn babies. Because the measurements are so accurate and the number of babies is so few, you are unlikely to get any babies of exactly the same weight, particularly in a small data set.

2.95, 2.72, 3.55, 2.68, 3.67, 3.22, 2.97, 4.22, 2.78, 3.86, 3.66, 4.01

CHAPTER 22 EQUATIONS

What Is an Equation?

An *equation* consists of two expressions linked by an equal sign (=).

Example

This is an equation: $13 - 2x = 5$

One expression is $13 - 2x$, and the other expression is just the number 5. The expression $13 - 2x$ consists of two terms 13 and $2x$. The letter x, stands for an unknown number. The term, $2x$, shows that x is multiplied by 2. The number 2 in the term $2x$ is called the coefficient.

As it stands, the equation, $13 - 2x = 5$, is neither true nor false, because the value of x is not specified.

If you replace x in the equation $13 - 2x = 5$ with the value 4, then you will have an equation that *is* true:

$$13 - 2 \bullet 4 = 5$$
$$13 - 8 = 5$$
$$5 = 5$$

Since it is true that $5 = 5$, you can say that $x = 4$ is a solution to the equation $13 - 2x = 5$.

What Does It Mean to Solve an Equation?

When you have an equation in terms of x, a solution of the equation is a value of x that makes both sides of the equation equal.

Example

The number 3 is a solution to the equation $4x + 6 = 18$ because $4 \bullet 3 + 6 = 18$. In the same way, the number 7 is not a solution to the equation $4x + 6 = 18$ because $4 \bullet 7 + 6 \neq 18$.

Solving an equation for an unknown number x means finding all of the values of x that make both sides of the equation equal.

How Many Solutions Can an Equation Have?

The answer to this question depends on the equation. Equations might have no solution, one solution, more than one, or even an infinite number of solutions.

An Equation with No Solution

Example

$6x + 10 = 6x + 20$

The equation, $6x + 10 = 6x + 20$, has no solution.

There is no value of x that will make both sides of this equation equal. This can be illustrated graphically by graphing $y = 6x + 10$ and $y = 6x + 20$ on the same axes.

You can see that the lines are parallel and never meet, thus they have no common point or no point of intersection.

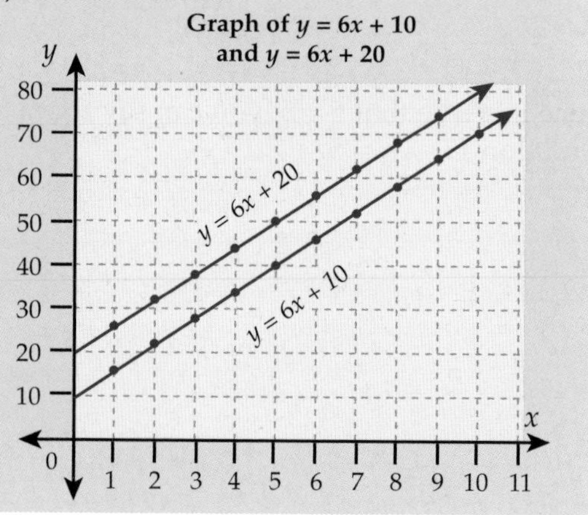

An Equation with One Solution

Example

$6x + 10 = 40 - 2x$

The equation $6x + 10 = 40 - 2x$ has one solution.

If x is replaced by 3.75 in the equation,

$$6x + 10 = 40 - 2x$$

then both sides of this equation are equal (to 32.5).

$x = 3.75$ is the only value of x that will make this equation true.

This can be illustrated graphically by graphing $y = 6x + 10$ and $y = 40 - 2x$ on the same axes.

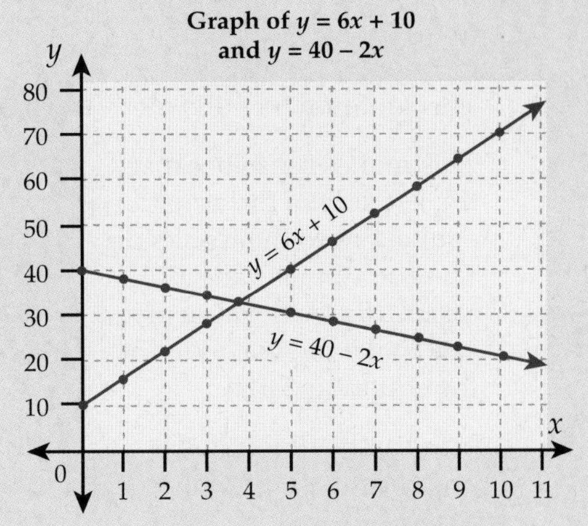

These lines intersect, and the solution is given by the x-value of the point of intersection of the two lines. The x-value of the point of intersection of the lines is between 3 and 4. This illustrates graphically that $6x + 10 = 40 - 2x$ has one solution. The solution can be estimated from the graph.

Chapter 22

An Equation Where Every Value of x Is a Solution

The equation $6x - 10 = 6(x - \frac{5}{3})$ is an example of an equation where every value of x is a solution.

Example

$6x - 10 = 6(x - \frac{5}{3})$

The equation $6x - 10 = 6(x - \frac{5}{3})$ has many solutions.

This makes sense because the left side of the equation is equivalent to the right side. The equation is called an *identity.* Any value of x will make the two sides of the equation equal.

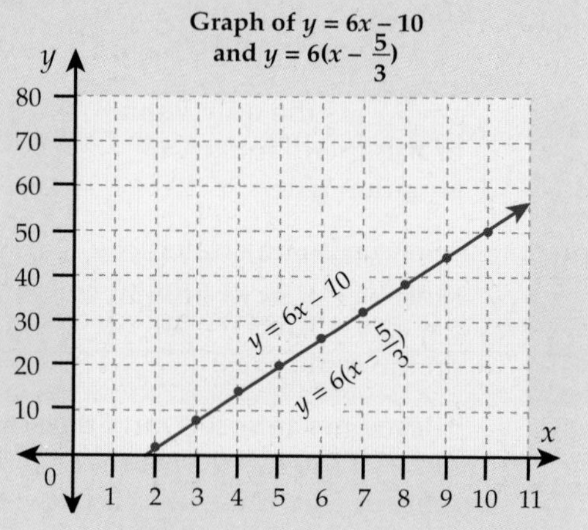

This can be illustrated graphically by graphing $y = 6x - 10$ and $y = 6(x - \frac{5}{3})$ on the same axes, and showing that both graphs lie along the same line or lie "on top of one another."

An Equation with Two Solutions

Example

The equation $x^2 = 1$ has two solutions.

Both 1 and –1 make the two sides of the equation equal. The two solutions to the equation are 1 and –1. This can be illustrated graphically by graphing $y = x^2$ and $y = 1$ on the same axes.

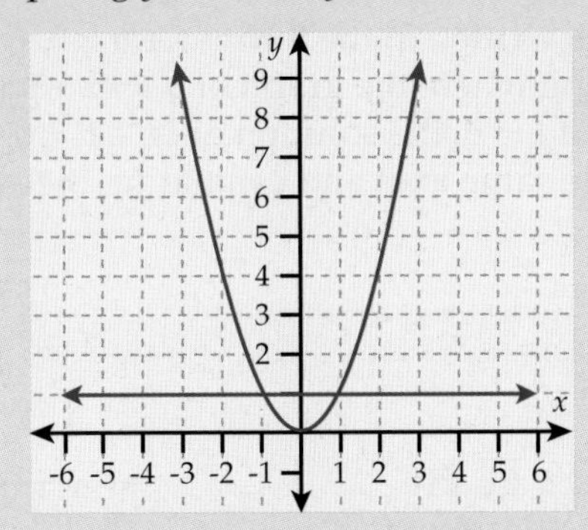

The line and the curve intersect in two places, and the x-values of the points of intersection are 1 and –1. This illustrates graphically that, $x^2 = 1$, has two solutions, 1 and –1.

Linear Equations

An equation is a linear equation in one unknown if it can be written in this form:

$$0 = kx + b$$

In $kx + b = 0$, the letters k and b are fixed numbers and x is an *unknown*. This means that x stands for some number or numbers that are not yet determined.

In a linear equation (expressed in terms of x), the power (or exponent) of x is 1.

If the power of x in any equation (expressed in terms of x) is raised to a power greater than or less than 1, the equation is not linear.

Example

The equation $x^2 = 9$ is not a linear equation.

When a linear equation is written in the form $kx + b = 0$, the solution of the equation can be found by estimating where the graph of $y = kx + b$ intersects the x-axis. This is called the x–intercept. If $y = 0$, the solution of $kx + b = 0$, can be found by graphing $y = kx + b$ and $y = 0$ on the same axes and estimating the x-value of their point of intersection.

Example

$0 = 2x - 10$

The graph of $y = 2x - 10$ is shown below. This graph intersects the x-axis (the line $y = 0$) at $x = 5$.

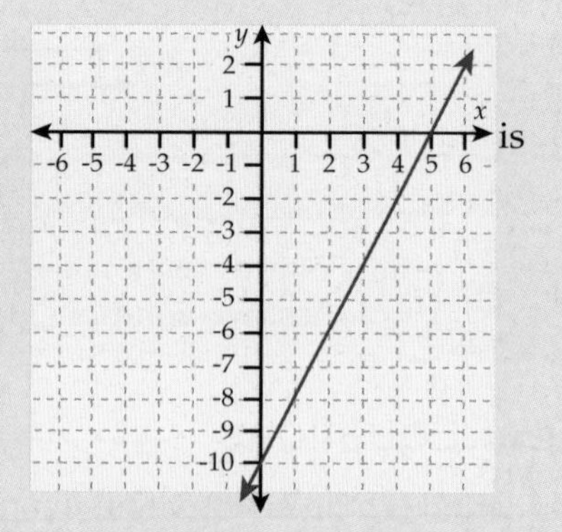

Check to see if 5 is a solution to the equation, $0 = 2x - 10$ by seeing if 5 will make both sides of the equation equal:

$0 = 2 \bullet 5 - 10$

$0 = 10 - 10$

$0 = 0$

So, yes, 5 is a solution to the equation, $0 = 2x - 10$.

Equivalent Equations Have the Same Solution

If you add the same number to both sides of an equation, the new equation is equivalent to the first equation and has the same solutions:

$$\text{If } a = b\text{, then } a + c = b + c.$$

If you subtract the same number from both sides of an equation, the new equation is equivalent to the first equation and has the same solutions:

$$\text{If } a = b\text{, then } a - c = b - c.$$

If both sides of an equation are multiplied by the same number (not equal to zero), the new equation is equivalent to the first equation and has the same solutions:

$$\text{If } a = b\text{, then } ac = bc.$$

If both sides of an equation are divided by the same number (not equal to zero), the new equation is equivalent to the first equation and has the same solutions:

$$\text{If } a = b \text{ and } c \neq 0\text{, then } \frac{a}{c} = \frac{b}{c}.$$

If the two sides of an equation are interchanged, the equation retains the same solutions.

$$\text{If } a = b\text{, then } b = a.$$

Balance Scale Illustration

A balance scale is often used to illustrate these properties. A scale that is in balance will remain balanced if two objects of the same weight are added to both sides or if two objects of the same weight are taken away from both sides.

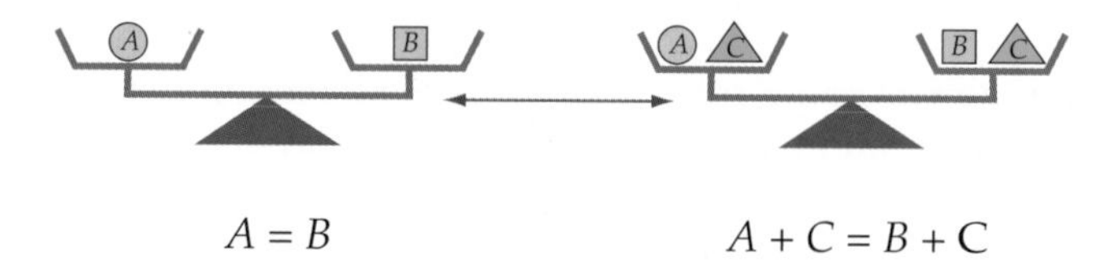

A balance scale will remain balanced when equal multiples of objects are added or are taken away from each side.

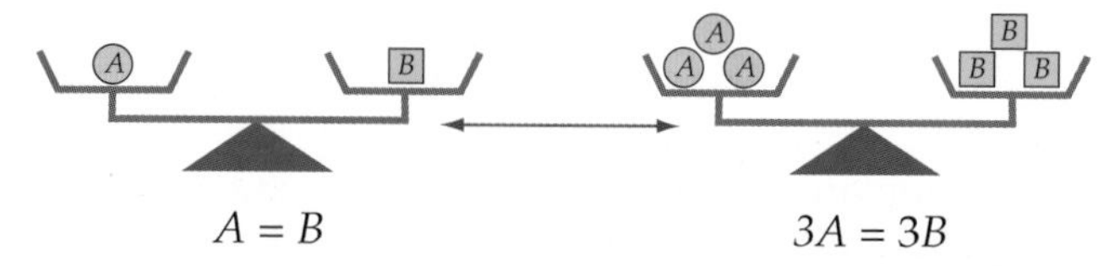

Solving Linear Equations in One Unknown

There are two main approaches to solving linear equations in one unknown—an algebraic approach and a graphing approach.

An Algebraic Approach to Solving Equations in One Unknown

This approach can be used to solve any linear equation in one unknown, and involves writing the equation that you are trying to solve as a simpler equivalent equation. Repeat this process until you arrive at an equation in which the solution is obvious. This solution is also the solution to the original equation.

Example

Solve the equation: $6x + 10 = 40 - 2x$.

First, write a simpler equivalent equation, such as, one with the unknown terms on the same side:

$$6x + 2x + 10 = 40 - 2x + 2x$$
$$8x + 10 = 40$$

Second, write another simpler equivalent equation, such as, one with the numerical terms on the same side:

$$8x + 10 - 10 = 40 - 10$$
$$8x = 30$$

If at this stage, you still cannot "see" the solution, write another simpler equivalent equation, such as one where the coefficient of x is 1:

$$\frac{8x}{8} = \frac{30}{8}$$
$$x = 3.75$$

Now the solution is obvious, it is 3.75. The equation, $x = 3.75$ is equivalent to the original equation, $6x + 10 = 40 - 2x$. The solution to the equation $x = 3.75$ is the same as the solution to the equation $6x + 10 = 40 - 2x$.

This algebraic approach can be used to solve any linear equation in one unknown. Each specific equation solution involves generating the equation's chain of equivalent equations, so the number of steps will differ. It is important to note that this method works for all equations.

Chapter 22

A Graphing Approach to Solving Equations in One Unknown

On page 351, you saw how to estimate the solution to the equation $6x + 10 = 40 - 2x$ by graphing the two linear equations in one unknown, $y = 6x + 10$ and $y = 40 - 2x$, on the same set of axes.

The two lines intersect at one point, and the solution is found by finding the x-value of the point of intersection of the two lines.

Any linear equation in one unknown can be solved using this method. The one drawback is that the solution must be estimated when you can't tell exactly what the x- and y-values are where the lines intersect.

Linear Equations

Linear equations can involve fractional or decimal coefficients, parentheses, and numerical terms on both sides of the equation. The algebraic method described provides a structured way of dealing with complex-looking equations.

Equations Involving Fractional or Decimal Coefficients

When an equation involves fractional or decimal coefficients, the equation can be converted into an equivalent equation that has integer coefficients.

Example

$$\frac{x+3}{6} - \frac{2x-3}{4} = 2$$

One way of converting this equation to an equivalent equation with integer coefficients is to multiply the left side and the right side of the equation by 12, giving this equivalent equation:

$$2(x + 3) - 3(2x - 3) = 24$$

Equations Involving Parentheses

When an equation involves parentheses, the equation can be converted into an equivalent equation that has no parentheses.

Example

$2(x + 3) - 3(2x - 3) = 24$

Written as an equivalent equation without parentheses:

$$2x + 6 - 6x + 9 = 24$$

Equations Involving Numerical Terms on Both Sides

When the equation that you are to solve has unknowns or numerical terms on both sides, the equation can be converted into an equivalent equation that has the unknowns on one side of the equation and numerical terms on the other side of the equation.

Example

$2x + 6 - 6x + 9 = 24$

This equation can be written as an equivalent equation with numerical terms on one side and terms with unknowns on the other:

$$\begin{aligned} 2x + 6 - 6x + 9 &= 24 \\ 2x + 6 - 6 - 6x + 9 - 9 &= 24 - 6 - 9 \\ 2x - 6x &= 9 \\ -4x &= 9 \\ x &= -\frac{9}{4} \end{aligned}$$

No matter how complex a linear equation may seem, it is important to remember that it can be solved by writing as many equivalent equations as needed to reveal the solution.

Chapter 22

Linear Equations in Two Unknowns

The equation $3x + y = 25$ is a linear equation in two unknowns. The solution to this linear equation in two unknowns is the set of all points (x, y) that satisfy the equation. For example, (4, 13) is one solution. All solutions lie on the line $y = -3x + 25$ with a slope = –3 and a y-intercept = 25. Two other solutions are (0, 25) and $(8\frac{1}{3}, 0)$.

The general form of an equation in two unknowns is $y = kx + b$, where x and y are the unknowns and k and b are constants that are fixed for any specific equation. The constant k gives the slope of the graph, and the constant b indicates where the graph crosses the y-axis, the y-intercept.

You may recognize that the slope of the line $y = 3x + 25$ is 3, and is given by the coefficient of the x-term (i.e., $k = 3$). You may also recognize that the line $y = 3x + 25$ crosses the y-axis at the point (0, 25) because the constant $b = 25$.

Solving Linear Equations in Two Unknowns

When you have two linear equations in two unknowns, you have what is called a "system" of equations. For such a system, the two linear equations can be graphed to find the solution. The coordinates of the point of intersection of the lines represented by the equations are the solutions to the system.

Suppose that you have these two equations in two unknowns:

Example

$3x + 5y = 11$
$7x - 9y = 5$

The graph of the system is:

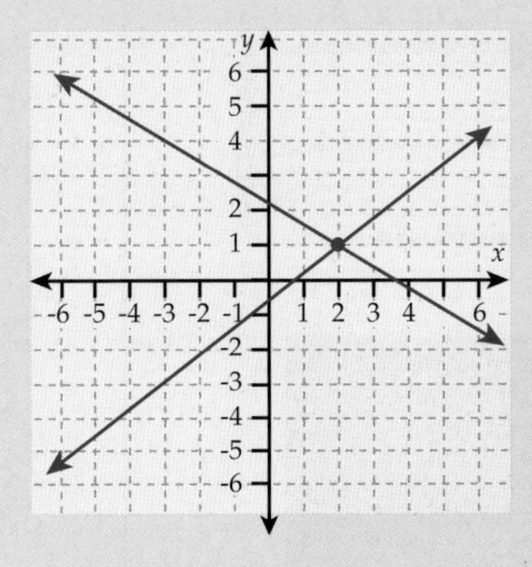

A solution to this system is by definition the coordinates of a point (x_1, y_1) that makes each equation true. That is, a single point that makes the right side equal to the left side in each equation.

This solution is the point on the graph at the intersection of the two lines, because this point lies on each line. In this case, the solution is $(x, y) = (2, 1)$.

From Linear Functions to Linear Equations

You have graphed linear functions (straight lines). The link between linear functions and linear equations will help you to understand linear equations more fully.

Example

Here, the books stacked on a 10-centimeter pallet are each 4 centimeters thick.

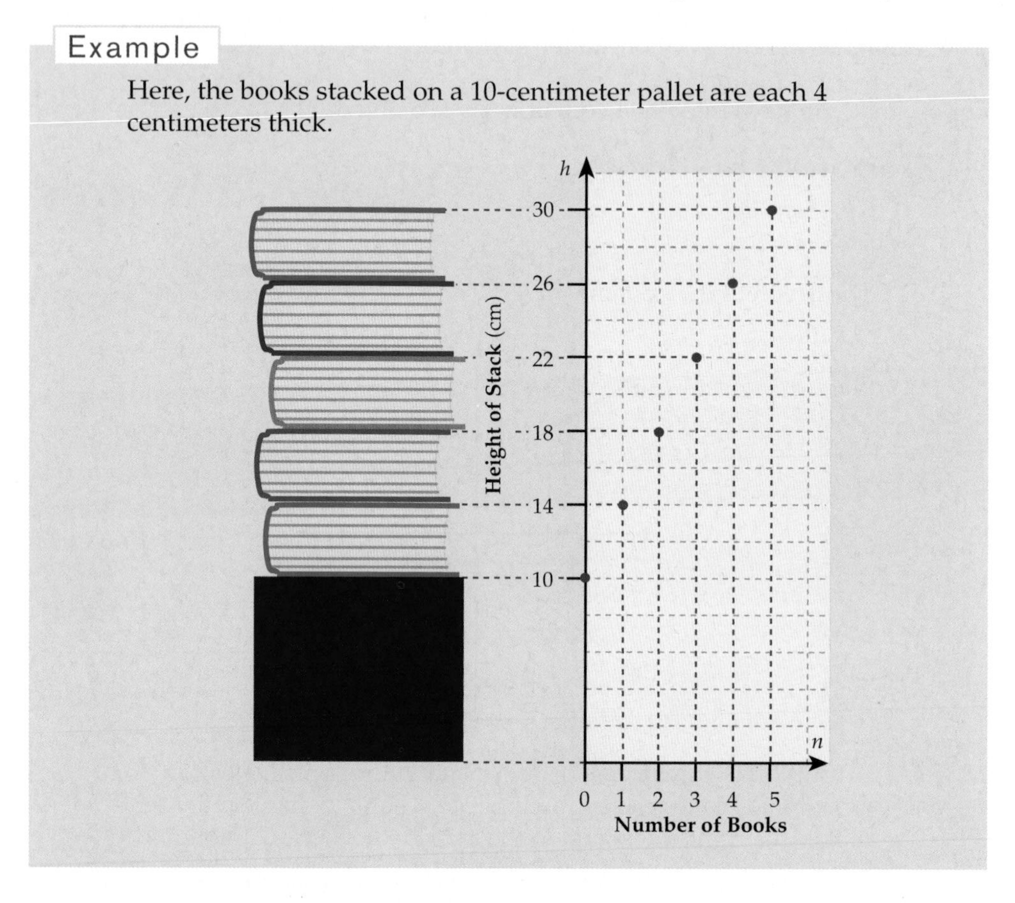

In this "stack-of-books" scenario, the height of the stack *depends* on the number of books in the stack. In mathematical terms, the height is a *function* of the number of books.

If you let the letter h represent the total height of the stack above the floor, and let the letter n represent the number of books in the stack, you could express h as a function of n:

$$h = 10 + 4n$$

This formula will allow you to answer some specific questions that you might have about this stack of books.

Example

How many books would it take to reach a height of 578 centimeters above the floor?

This means that for some unknown number of books, $h = 578$.

The problem is to find the value of the input variable (the number of books) that will give the output (the height of the stack of books) of 578 centimeters.

You are given that $h = 578$, but you also know that $h = 10 + 4n$. This means $578 = 10 + 4n$.

$$578 = 10 + 4n$$

$$578 - 10 = 10 - 10 + 4n$$ Subtract 10 from each side.

$$568 = 4n$$

$$\frac{568}{4} = \frac{4n}{4}$$ Divide each side by 4.

$$142 = n$$

A stack of 142 books would have a height of 578 centimeters.

Chapter 22

Representing Word Problems with Linear Equations

The challenge in a word problem is usually writing an equation to represent the given information.

The only way to master this aspect of linear equations is to try to understand the meaning of the word problem.

Once the meaning is understood, try to express the relation between the quantities in the equation. Remember that an equation consists of two expressions connected by an equal sign.

Then, solve the equation to answer the word problem. (See Chapter 3, *Using Algebra to Solve Problems.*)

Index

Symbols

A

B

C

D

E

F

N

O

P

Q

R

S

T

U

V